# Molecular Dynamics of Nanostructures and Nanoionics

# Molecular Dynamics of Nanostructures and Nanoionics

## Simulations in Complex Systems

**Junko Habasaki**

JENNY STANFORD
PUBLISHING

*Published by*

Jenny Stanford Publishing Pte. Ltd.
Level 34, Centennial Tower
3 Temasek Avenue
Singapore 039190

Email: editorial@jennystanford.com
Web: www.jennystanford.com

**British Library Cataloguing-in-Publication Data**
A catalogue record for this book is available from the British Library.

**Molecular Dynamics of Nanostructures and Nanoionics: Simulations in Complex Systems**



ISBN 978-981-4800-77-8 (Hardcover)
ISBN 978-1-003-04490-1 (eBook)

# Contents

# Preface

Nanostructured materials with several components and complex structures have drawn researchers' attention and are expected to develop further for material designs in many applications in electrochemical, colloidal, medical, and pharmaceutical fields but, not limited to these. This book, *Molecular Dynamics of Nanostructures and Nanoionics: Simulations in Complex Systems,* discusses complex nanostructured systems exemplified by nanoporous silicates, spontaneously formed gels from silica-nanocolloidal solutions and related systems examined by molecular dynamics (MD) simulations [1–4]. Nanoporous materials, nanocolloidal systems, and gels are useful in many applications, such as electric devices, storage and delivery of gas and ions, and drug delivery.

This book gives an overview of the history, current status, and frontiers of the field. It discusses the fundamental aspects related to the common behaviors of some of these systems and common analytical methods to treat them. The book has been written for researchers, newcomers, and students and will be of great interest to condensed matter physicists, chemists, materials scientists, and engineers.

After a general introduction to classifications, the effects of nanosizing and historical views for nanoionics are explained. The focus is on the works by classical MD simulations, which play an important role in examining a large system, which may have long relaxation times. *Ab initio* MD simulations such as the Car–Parrinello method [5] are also useful and have played important roles in recent studies and are expected to gain importance for further examination of complex systems in the future. Therefore, research by these methods will be referred to from time to time. Unfortunately, in contemporary conditions, *ab initio* MD simulations may have some limitations for system sizes and time scales, which can be treated and some cautions may be required to understand the results obtained by them. Common problems with classical MD to treat complex systems have been discussed in this book.

The motivation of writing this book is as follows. While writing previous our book for ionics [6], I strongly felt a need for a better understanding of more fundamental aspects of nanostructured systems. That is why I started several projects concerned with these problems, and therefore, the key problems treated in this book are enhanced ion dynamics in nanoporous systems [7–9] and self-organization of gels in the colloidal systems examined by us [10].

Further developments of these fields are expected in order to design better nanofunctional materials such as self-healing systems, drug delivery systems, catalysts, and nanomachines. Undoubtedly, MD simulations can contribute to such developments, because they can treat problems from several different kinds of approaches, including ones that are difficult to be evaluated by experimental methods. MD simulations' role is not limited to reproducing the phenomena; they also help to understand these phenomena and make further predictions.

On the frontiers of material sciences, simulations of these complex systems have been already used in many systems and successful results have been obtained. Some of these results have been presented in this book. However, a fundamental understanding based on common behaviors of these systems seems to be still lacking and awaits further developments by researchers and some common concepts and tools to treat them are required. In this book, after explaining the essential part of the fundamentals of MD simulations, common tools to examine structures is introduced for characterizing nanostructured materials. They include fractal (and multifractal) dimension analyses, characterization of networks by $Q_\mathrm{n}$ structures, and length scales determined from charge density waves and density waves. The contents of this book are as follows.

First, the classification of nanostructured materials is expanded for the inclusion of disordered and/or porous materials. Then, general arguments for the effects of nanosizing are given.

Second, historical views and problems to be solved in some of nanostructured systems are shown. Experimentally, enhanced dynamics of ions of nanocomposites were found in LiI-$Al_2O_3$ [11] and in related materials. Especially, the book discusses the mechanism of enhanced diffusion (or electric conductivity) found in composites and porous ionic systems. It introduces experimental works as well as some MD works along with the history of research on these systems. It can be a good guide for understanding the problems to be treated.

Third, the effects of introducing pores are examined to learn how the dynamics of the system are modified by additional factors. The discussion includes how the possible mechanisms in the enhancement of Li ion conduction in porous lithium silicates can be distinguished by MD simulations. The changes in time scales of events of different length scales are examined in order to understand what mechanism is responsible for the enhancement. Dynamics of such structures may show heterogeneous dynamics with coexistence of fast and slow diffusive motions in certain time scales and temperatures. Dynamic heterogeneity in porous materials is also examined.

Fourth, in spite of recent developments in the field, our understanding of colloidal systems seems to be rather limited, because of many components and complex structures found in these systems. Colloidal solution treated in full atomistic simulations can open new frontiers for this problem, although such approaches are challenging because of the large difference in the size of colloidal units and solvent molecules. In this book, the mechanism of coagulation with different salt concentrations is examined by full atomistic simulations of the silica nanocolloidal solution with salt. Structural changes characterizing systems are found not only for the silica network, but for the solvent and salt.

Barrier heights of coagulation processes and structures of gels thus formed are examined and discussed in this book. The structures obtained by MD simulations of several colloidal systems are compared, and fractal dimensions are found to be useful to characterize them. It is a basis of the new classification of nanostructured materials proposed in Chapter 2.

This book is useful for introducing the current status and perspectives for applying MD simulations for nanostructured systems and can be a guide to select the suitable methods to simulate such systems.

The book is inspired by our previous book on ionics that covered theories, experiments, and simulations. The author, would like to thank the coauthors of the previous book, Carlos León and Kia L. Ngai.

**Junko Habasaki**

## References


1. Alder, B. J., Wainwright, T. E. (1959), Studies in molecular dynamics. I. General method *J. Chem. Phys.* 31, pp. 459–466; https://doi.org/10.1063/1.1730376.
2. Alder, B. J., Wainwright, T. E. (1960), Studies in molecular dynamics. II. Behavior of a small number of elastic spheres, *J. Chem. Phys.* 33, pp. 1439–1451. https://doi.org/10.1063/1.1731425.
3. Woodcock, L. V. (1971), Isothermal molecular dynamics calculations for liquid salts, *Chem. Phys. Lett.*, 10, pp. 25–261.
4. Verlet, L. (1967), Computer "experiments" on classical fluids. I. Thermodynamical properties of Lennard–Jones molecules, *Phys. Rev.*, 159, pp. 98–103.
5. Car, R., Parrinello, M. (1985), Unified approach for molecular dynamics and density-functional theory, *Phys. Rev. Lett.*, 55, pp. 2471–2474.
6. Habasaki, J., León, C., Ngai, K. L. (2017), *Dynamics of Glassy, Crystalline and Liquid Ionic Conductors: Experiments, Theories, Simulations*, Springer International.
7. Habasaki, J. (2016), Molecular dynamics study of nano-porous materials—Enhancement of mobility of Li ions in lithium disilicate in NVE conditions, *J. Chem. Phys.*, 145, pp. 204503.
8. Habasaki, J., Ngai, K. L. (2018), Heterogeneous dynamics in nano porous materials examined by molecular dynamics simulations: Effects of modification of caged ion dynamics, *J. Non-Cryst. Solid*, 498, pp. 364–371.
9. Habasaki, J., Ngai, K. L. (2017), nearly constant loss (NCL) in lithium metasilicate glass at low temperatures—anisotropic and dynamical caging from molecular dynamics simulations, *J. Phys. Chem. C*, 121 (25), pp 13729–13737.
10. Habasaki, J., Ishikawa, M. (2014), Molecular dynamics study of coagulation in silica-nanocolloid–water–NaCl systems based on the atomistic model, *Phys. Chem. Chem. Phys.*, 16, pp. 24000–24017.
11. Liang, C. C. (1973). Conduction characteristics of the lithium iodide–aluminum oxide solid electrolytes, *J. Electrochem. Soc.*, 120(10): pp. 1289-1292; doi:10.1149/1.2403248.
12. Kirkwood, J. G., (1935), Statistical mechanics of fluid mixtures, *J. Chem. Phys.*, 3, pp. 300–313.

## Chapter 1

# Introduction to Molecular Dynamics Simulations of Complex Systems

In recent developments of nanotechnology, nanostructured systems formed by mixing or coexistence of several constituents and substructures are often the target of debates. Nanostructured materials treated in this book include porous materials, gels, and some composites, where the structures at the nanoscale take important roles to exhibit their properties.

For the development of nanotechnology, experiments, theories, and simulations have their roles and researchers of these methods are expected to work hand in hand. In previous our book for ionics [1], problems have been treated in such a manner. We have treated recent progress of nanoionics there as well as ionics in fundamental systems. During that project, as one of the researchers of molecular dynamics simulations, I felt that deep insights from a molecular or atomistic level of arguments are still not adequate in the field of nanoionics and related ones, although many experimental facts and theories and/or explanations were so far proposed. That is why we started several projects including enhanced dynamics in porous materials and fully atomistic simulations of colloidal systems. In these fields,

*Molecular Dynamics of Nanostructures and Nanoionics: Simulations in Complex Systems*
Junko Habasaki

ISBN 978-981-4800-77-8 (Hardcover), 978-1-003-04490-1 (eBook)
www.jennystanford.com

simulations are expected to add new insights and to bridge the gap between theory and experiments. Furthermore, new directions for works may be suggested by them. In the present book, I would like to introduce interesting problems in nanoionics and hints to solve it with examples of some MD simulations. Before mentioning the results of these trials, some fundamentals, tools, a brief history, and current status of the fields will be discussed.

The system with modified nanostructures such as nanoparticles, nanoporous materials, and nanofilms often exhibits properties that are drastically different from the crystalline and/or conventional polycrystalline materials. Such properties tend to be explained by large surface areas (surface and volume ratio) or by the existence of grain boundaries. Of course, for some of the properties, the large surface area will play significant roles (see Chapter 2). However, question arises if the effects of the surface can explain all the properties of nanostructured materials. If one traced nearly a half-century researches for enhanced ion dynamics in such nanostructured materials, one would find that researchers have proposed many models and theories to explain phenomena until nowadays [1]. This fact itself means that the explanation by the surface and/or boundary alone is not necessarily satisfactory. Even a role of surface and boundary is not necessarily clarified yet, although there are good and/or promising theories and models explaining some of these features successfully. To understand problems and possible origins, a brief history of nanoionics will be examined in Chapter 3.

In the present book, we focus on two key factors to affect structure and dynamics. It was found that introducing pores can cause enhanced dynamics of ions in the MD simulations of nanoporous silicate under constant volume conditions. One important factor discussed in this book is the change in caging dynamics in such nanostructured materials with introducing pores.

Modification of structures caused by the addition of salt to the colloidal solutions will be examined and the coexistence of different length scales in structures will be shown for colloidal solutions, especially in gels. Tools to measure length scales, fractal dimension, $d_f$, and to characterize networks will be introduced to examine them. Then, another key factor is concerned with fractality [2, 3] of the systems and coexistence of different length scale of the system, that is multifractality [4–6]. The change of fractal dimension observed is closely related to the length scale related to the size of the particles, domains, and other substructures. These factors also modify dynamics, since the cooperatively moving regions of particles (ions, molecules) might be cut in such structures. In other words, in nanostructured materials, longer length scale motions related to the long-ranged structures are missing or modified.

These characteristics of structures as well as dynamics are important for not only fundamental studies but also applications in many fields. Related to dynamics in porous systems and length scales, it is interesting to consider the following phenomena. A phase-diagram and related properties such as the glass transition temperature are known to be modified, when the system is trapped in pores or films. Such nanostructured materials can be used in drug delivery systems [7], and changes in the glass transition temperature of drugs in porous systems are used to control stability and/or releasing power.

Systems with complexity treated by simulations are growing in numbers rapidly; however, it seems to be still in developing stages. It is noteworthy that even for the single-component system, such as the soft-sphere or Lennard–Jones model, the behavior of systems is sometimes a quite complex one including heterogeneous dynamics with several length scales; despite developments of experiments, theories, and simulations, a mechanism of "the glass transition" is often regarded as an unsolved problem [1, 8–12], including these simple systems. Interestingly, this kind of complexity is also observed in ionics [1], colloidal systems [13], and biosystems [14], where heterogeneous dynamics and multifractal structures suggesting the existence of deterministic dynamics [1, 12] are commonly found. Therefore, it is useful to argue situations of simple ionic systems such as lithium silicate glass; then, it will be compared with the system with modifications.

To perform the above-mentioned tasks, choosing suitable simulation methods to examine nanostructured materials is an important problem. Computational chemistry and physics have been used in these interesting fields of nanostructured materials as well as nanoionics. At the beginning of computational science, 3M (molecular mechanics (MM), Monte Carlo (MC), and molecular dynamics (MD)) were often regarded as the principal methods. Nowadays, many methods and their hybrids are included in this category. Density functional theory became a useful and conventional method for quantum chemistry as well as molecular orbital methods and are often used complementary to simulations. Moreover, *ab initio* MD such as the Car-Parrinello method [15] became popular. Of course, this kind of approach is useful if the suitable conditions are carefully chosen especially for the length and time scales to be covered; however, it is not necessarily easy to cover them, especially for complex systems. Furthermore, systems used for applications often have several different length scales, domains, and substructures formed by many constituents and therefore, direct modeling of such systems is not necessarily easy. Under these circumstances, classical MD simulations using potential parameters based on *ab initio* quantum chemical calculations are reasonable approaches for many purposes, because this can connect electrostatic and atomistic-level approaches. In the *ab initio* approaches, equilibrated structures by classical MD are often necessary as inputs to save an equilibration time.

In typical cases, system size which can be covered by the classical MD is from several hundred to several ten thousand particles, while for examining nanostructured materials, the system size required will be comparable to it or larger due to the existence of domains with different sizes and/or hierarchy of structures as well as existence of long-ranged structures. On the other hand, if the system size becomes too large, it is difficult to perform calculations within reasonable calculation time. Therefore, the choice of a suitable system size is important for the setting of MD simulations. Meanwhile, the time scale scale covered by MD simulations is typically from several hundred ps to several ns. For some purposes, it may be too short for examining the system especially at low-temperature and high-pressure regions. This is because the relaxation time becomes longer and longer with decreasing temperature and increasing pressure in most cases. Therefore, careful choice of the combination of system sizes and time scales with considering dynamics is required.

In many research fields such as physics of disordered materials, biochemistry, environmental chemistry, and material chemistry, one may encounter the problem of slow dynamics, as already mentioned where extremely heterogeneous dynamics are found. The problem of the sampling of rare events and the existence of heterogeneity of the motions need to be considered for statistical treatment of the system. Therefore, setting of reasonable time scales to cover such events and different stages caused by such motions is another requirement for reliable simulations. Examples in the present work give a hint on how to choose them.

For processing the above-mentioned discussions effectively, the organization of this book is planned as follows.

In Chapter 2, first, the classification of nanostructured materials found in the literature is summarized. The classification using the dimensionality of the systems is found to be useful for many purposes. However, the system treated in the present book is not limited to crystals or polycrystals but includes disordered materials and it is not easy to classify gels and/or porous materials by these classification methods. Thereby, classifications based on the dimensionality of nanostructures are expanded to fractal [2] and multifractal [3–6] ones. Possible methods to classify porous materials based on dimensionality are also proposed. Expanded indexing can be used not only to classify known systems but also to design new nanostructured materials.

The effects of nanosizing so far suggested, such as large surface area (or region), and some of the related theories are briefly explained here, although these effects are not necessarily sufficient to explain the properties of nanostructured systems.

In Chapter 3, the historical view for some complex systems, which show enhanced dynamics will be shown with experimental observations and possible explanations. To consider other controlling factors of dynamics and structures in nanostructured materials, a history of research in nanoionics will be useful. In 1973, Liang [16] reported enhanced dynamics of ions in the lithium iodide-aluminum oxide system. After that, this and related systems have been examined by many researchers. These works brought developments in studies of "surfaces and boundaries," "chemical preparations of composites," as well as developments of "percolation theories." Therefore, the results of these studies will be useful to learn the history and current status of the field of nanoionics. This can enumerate factors related to the properties of complex systems and suggest what is necessary for the development of the field.

In Chapter 4, fundamentals of molecular dynamics to treat such problems are introduced with tools utilized to analyze structures and dynamics of systems after explanations of the essence of methods. In such materials, one may need to examine the coexistence of several length scales and/or networks. Therefore, the method to examine length scales and network systems will be explained. Concepts of fractal and multifractal useful to characterize nanostructured systems are explained as well. MD simulations and their application to several fields can be found in other books [1, 17, 18].

In Chapter 5, time development of the mean squared displacement (MSD) will be summarized for the case of $Li^+$ ions in lithium silicate systems [19, 20]. Several stages of dynamics found in ionics, which start from caging (nearly constant loss, NCL) region to a diffusive regime, through a regime of jump motions (independent and cooperative jumps), are observed and it was clarified what kind of motions contribute to complex behaviors in each stage. It also explains that found in the frequency dependence of conductivity spectra. This is a representative example of time (or frequency) dependent dynamics, where the method can cover the wide time ranges from ~fs to several ~tens of ns. Thus, MD simulations enable us to catch the phenomena of wide ranges of time scales consistently and reasonably. A time scale (and a length scale related it) of each event observed there can be used as a basis of understanding the effects by modification of systems found in porous systems [21, 22] shown in the following chapters.

In Chapter 6, enhanced dynamics in constant energy conditions (*NVE*) in porous lithium disilicate will be discussed. In this case, pores are introduced by the expansion of the system in *NVT* conditions. Enhanced dynamics with the maximum at a certain density (or porosity) are predicted there. In that chapter, possible causes of enhancements are separated and the importance of changes in caged ion dynamics for enhancement is shown. It can be a specific example to consider how modification of the structure affects dynamics.

In Chapter 7, the effects of introducing pores in lithium disilicate systems were examined further to understand how heterogeneous dynamics are modified by them. It will be useful to understand the existence of common behaviors in many kinds of systems as discussed in ref. [1]. Self-organization of new structure with larger voids in the porous system in *NVE* conditions is found in MD simulations and this process can be explained as the stress-relaxation. Interestingly, even for the system with large voids obtained in *NVE* runs, healing processes are found in the subsequent MD simulations in *NPT* conditions. They are not just simple shrinks by pressure but are reformation processes of coordinated structures with reconnection of bonds. Repairing of the system tends to occur before attaining a diffusive process, in the case of small expansion region, while the contribution of the thermal activation process is observed in the healing starting from low density systems. The remarkable repairing is found in metasilicate with a larger amount of Li ions.

In Chapter 8, results of atomistic simulations of the nano-colloidal silica systems mixed with water and salt, which show spontaneous aggregation and gelation phenomena, are discussed. The process is also provided the method of preparation of natural porous systems.

Usually, to perform reliable MD simulation, large system size and long runs with enough equilibration time are required. Unfortunately, it is a time-consuming and costly work. This is problematic when the simulations of colloidal systems are planned. By results in the present chapter as well as experimental results introduced in Chapter 3, one can understand how the existence of different length scales is concerned with the formation of the gel. The importance of the concept of the potential of mean force [23] for the formation of the gel will be shown and differences from the DLVO model [24, 25] will be argued.

When the gel is formed, structures of water are changed from low density to high density, and this change is related to the polyamorphism of water [26].

In Chapter 9, structures of the gel thus spontaneously formed from atomistic simulations are examined in detail. Complexity discussed for this example is concerned with fractal dimensions of network structures and structural changes in the solvent during gelation. As found in these examples, multifractality of structures is one of the key features of the nanostructured systems. The system can be understood as multiple networks of silica, water, and salt parts.

In the afterword, perspectives for the studies in these fields are shown with a summary of the previous chapters. It is useful to point out that common features for these examples can be concerned with multifractal structures with the coexistence of several length scales and several characteristic times related to these length scales. Since research of these systems is not completed but is in progress, this book can be a useful introduction for new researchers for these interesting fields.

In the field of nanostructured and complex materials, molecular dynamics simulation will play important roles. I hope that the introduction to this field by this book will be beneficial to researchers and students, and it will promote productive collaborations of researchers.

## References

1. Habasaki, J., León, C., Ngai, K. L. (2017), *Dynamics of Glassy, Crystalline and Liquid Ionic Conductors: Experiments, Theories, Simulations*, Springer International.
2. Mandelbrot, B. (1967), How long is the coast of Britain? statistical self-similarity and fractional dimension, *Science*, 156, pp. 636–637.
3. Ben-Avraham, D., Havlin, S. (2000), *Diffusion and Reactions in Fractals and Disordered Systems*, Cambridge University Press, Cambridge.
4. Mandelbrot, B. (1999), *Multifractals and 1/f Noise: Wild Self-Affinity in Physics (1963–1976)*, Selecta Volume N. Springer.
5. Chhabra, A., Jensen, R. V. (1989), Direct determination of the f(alpha) singularity spectrum, *Phys. Rev. Lett.*, 62, pp. 1327–1330.
6. Habasaki, J., Ngai, K. L. (2011), Multifractal nature of heterogeneous dynamics and structures in glass forming ionic liquids, *J. Non-Cryst. Solids*, 357, pp. 446–453.
7. Kamaly, N., Yameen, B., Wu, J., Farokhzad, O. C. (2016), Degradable controlled-release polymers and polymeric nanoparticles: Mechanisms of controlling drug release, *Chem. Rev.*, 116(4), pp. 2602–2663.
8. Angell, C. A. (1988), Perspective on the glass transition, *J. Phys. Chem. Solids*, 49, pp. 863–871.

9. Ngai, K. L. (2007), Why the glass transition problem remains unsolved? *J. Non-Cryst. Solids,* 353 (8–10), pp. 709–718.

10. Biroli, G., Garrahan, J. P. (2013), Perspective: The glass transition, *J. Chem. Phys.,* 138, 12A301 (1–13).

11. Ngai, K. L., Prevosto, D., Capaccioli, S., Roland, C. M. (2008), Guides to solving the glass transition problem, *J. Phys: Cond. Matter,* 20, 244125(1–9).

12. Ngai, K. L. (2011), *Relaxation and Diffusion in Complex Systems,* Springer, New York.

13. Weeks, E. R., Crocker, J. C., Levitt, A. C., Schofield, A., Weitz, D. A. (2000), Three-dimensional direct imaging of structural relaxation near the colloidal glass transition, *Science,* 287(5453), pp. 627–631, DOI: 10.1126/science.287.5453.627.

14. Losa, G. A., Merlini, D., Nonnenmacher, T. F., Weibel, E. R. (2005), *Fractals in Biology and Medicine,* vol. 4, Springer Science & Business Media.

15. Car, R., Parrinello, M. (1985), Unified approach for molecular dynamics and density-functional theory, *Phys. Rev. Lett.,* 55, pp. 2471–2474.

16. Liang, C. C. (1973), Conduction characteristics of the lithium iodide-aluminum oxide solid electrolytes, *J. Electrochem. Soc.* 120(10), pp. 1289–1292, doi:10.1149/1.2403248.

17. Allen, M. P., Tildesley, D. J. (2017), *Computer Simulation of Liquids 2nd ed.,* Oxford University Press.

18. Edited by Massobrio, C., Du, J., Bernasconi, M., Salmon, P. S. (2015), *Molecular Dynamics Simulations of Disordered Materials from Network Glasses to Phase-Change Memory Alloys,* Springer.

19. Habasaki, J., Hiwatari, Y. (1995), Origins of the two-step relaxation and the boson peak in an alkali silicate glass studied by molecular-dynamics simulation, *Phys. Rev. E,* 52, 2681.

20. Habasaki, J., Ngai, K. L., Hiwatari, Y. (2002), Molecular dynamics study of cage decay, near constant loss, and crossover to cooperative ion hopping in lithium metasilicate, *Phys. Rev. E,* 66, 021205(1–11).

21. Habasaki, J. (2016), Molecular dynamics study of nano-porous materials—enhancement of mobility of Li ions in lithium disilicate in NVE conditions, *J. Chem. Phys.,* **145**, p. 204503.

22. Habasaki, J., Ngai, K. L. (2018), Heterogeneous dynamics in nano porous materials examined by molecular dynamics simulations—effects of modification of caged ion dynamics, *J. Non-Cryst. Solid.,* 498, pp. 364–371.

23. Kirkwood, J. G. (1935), Statistical mechanics of fluid mixtures, *J. Chem. Phys.*, 3, pp. 300–313.
24. Derjaguin, B., Landau, L. (1941), Theory of the stability of strongly charged lyophobic sols and of the adhesion of strongly charged particles in solutions of electrolytes, *Acta Phys. Chem. URSS*, 14, pp. 633–662.
25. Verwey, E. J. W., Overbeek, J. Th. G. (1948), *Theory of the Stability of Lyophobic Colloids*, Amsterdam, Elsevier.
26. Mishima, O., Calvert, D. L., Whalley, E. (1985), An apparently first-order transition between two amorphous phases of ice induced by pressure, *Nature*, 314, pp. 76–78.

# Chapter 2

# Classification of Nanostructured Materials and Effects of Nano-Sizing

Nanostructured materials (NSM) have useful properties such as high conductivity, high storage power of materials and energy if a suitable combination of constituents and nanoscopic structures are designed. Especially, enhanced dynamics found in nanoionics [1] is an important issue to be discussed.

In this book, complex systems such as nanoporous materials [2, 3] and colloidal solutions [4] are examined by molecular dynamics (MD) simulations. Before explaining the results of these works, some preparations for the treatment of these systems are necessary. First, the classification of the nanostructured materials is discussed in this chapter and it will be expanded to cover fractal and porous structures based on our several projects concerning complex systems. Then the general effects of nano-sizing will be discussed. Here the term "fractal" does not necessarily mean the exact self-similar shapes, but it includes statistical ones. Details of the concept of the fractal will be explained later in Chapter 4.

*Molecular Dynamics of Nanostructures and Nanoionics: Simulations in Complex Systems*
Junko Habasaki

ISBN 978-981-4800-77-8 (Hardcover), 978-1-003-04490-1 (eBook)
www.jennystanford.com

## 2.1 What Are Nanostructured Materials?

Definitions of nanostructured materials are not the same if different researchers and fields are concerned. First, some methods found in the literature will be explained.

### 2.1.1 Classifications of Nanostructured Materials Found in the Literature

In 1995, Gleiter called the materials with a nanometer-sized (typically 1–10 nm) microstructure "nanostructured materials (NSMs)" [5, 6], where the structure elements discussed are mostly crystallites. In these articles, 12 typical structures are classified by combinations of four families (different chemical composition of crystallites) and three categories (different shapes of crystallites-layer-shaped, rod-shaped, and equiaxed crystallites).

Later, Pokropivny and Skorokhod [7] suggested classification based on the dimensionality of nanostructures themselves ($k$ = 0,1,2,3) and of their components ($l$, $m$, $n$ = 0,1,2,3) and a total of 36 classes are shown, with restricting the classification by 5 main ternary structures of kDlmn type built of 3 sort units. The system size $L < 100{\sim}500$ nm is considered there. For example, No. 35 (3D320) is for powder-layers composites.

As discussed in ref. [7], a general dependence of density of electron state (DOS) on the dimension of the nanostructure is well known. Namely, $\rho(E){\sim}\sqrt{E}$, $\rho(E)$ = const., $\rho(E){\sim}1/\sqrt{(E{-}E_0)}$, and $\rho(E){\sim}\delta(E{-}E_0)$ are for the 3-D, 2-D, 1-D and 0-D nanostructures, respectively. In general, diffusivity and conductivity also depend on the dimensionality as found in the Einstein equations. The dimension of the structure is also closely related to the surface area or related characteristics of nanoscale systems as discussed later. Thereby, similar classifications using dimensionality were used by several authors.

In 2011, Liu et al. [8] discussed the heterogeneous nanostructured electrode, where the nanostructured materials were also classified based on their structural complexity. That is, systems were classified from zero-dimensional (0-D) to three-dimensional (3-D). For example, 0-D includes core/shell (or core–shell) [9–11] nanoparticles, one-dimensional (1-D) includes coaxial nanowires, two-dimensional (2-D) includes graphene-based composites and three-dimensional (3-D) includes mesoporous carbon-based composites and the even more complex hierarchical 3-D nanostructured networks.

These classifications can treat "known systems" successfully and will be useful for many purposes, although suitable classification may depend on each purpose.

For example, by classification, one may have a clear picture of materials.

In the following sections, a similar scheme to classify complex nanostructured materials will be introduced. Although it depends on the same basis of dimensionality of nanostructures as in [7, 8], the combinations will not be restricted to the limited number of combinations and the classification will be expanded to further modification of systems.

The purpose of this trial is three-fold.

1. Complicated NSMs tend to have several components, several structural units with different fractal dimensions [12–16] and length scales, different porosity (porosity is usually defined as a fraction of pores or voids in the material). They may have different physical and chemical properties such as colors, mechanical properties, heterogeneities and so on. Possible combinations of them seem to be infinite. Therefore, the expansion of the classification to include more systems with different combinations seems to be required, so that one can find out new combinations for designing materials. Accordingly, classification, which can include many unknown systems without the restriction of classifications, is desirable.
2. In this book, ionically conducting glasses and colloidal solutions and its coagulation, as well as other complex systems, are examined. Furthermore, the concept of multifractal [13–16] is introduced because more than one characteristic region of the length scale is found in colloidal systems and gels [4, 17, 18].

   Therefore, the classification of systems is better to be extended to use fractal dimensions, to treat disordered and/or fractal systems.

3. There are several methods to treat porous systems. Some of them are discussed here. One of the possible methods is to use the apparent dimension for typical porous systems.

   An approach from a different direction is to use packing factor (or porosity) or fractal dimensions as parameters to control and characterize properties such as conductivity and mechanical properties.

   The classification using these values can treat the porous systems systematically. In Chapters 6 and 7, it will be shown that the diffusivity (and hence conductivity) of ions is modified considerably by changes in the porosity.

## 2.2 Classification of Substructures by the Dimension of Elemental Structural Units and Dimension of Space

In this chapter, a classification based on the dimensionality will be extended for mixtures, fractal structures, porous structures, and dispersed structures. The outline of explanations of these concepts is shown in Fig. 2.1.

### 2.2.1 Elemental Units of Nanostructures

In ref. [7], nanostructures are distinguished from the nanostructured materials, which can be bulk. Similarly, we consider the whole structure of nanostructured material and substructures of them separately. Substructures are formed by the elemental nanostructural units shown in Fig. 2.2.

For useful classifications, one needs to choose the elemental unit and the length $\lambda$ to characterize the nanostructured materials, where $\lambda$ is a length scale determining the range of sizes to treat as nanomaterials. It can be typically 10–100 nm; however, there are some differences among researchers. If it is compared with the characteristic length scale of the phenomena, a clear definition of it will be required. If no description of $\lambda$ is given, the natural way is to choose $\lambda$ as 1–1000 nm exactly stated in the term "nanostructured materials."

OUTLINE OF CLASSIFICATION METHODS

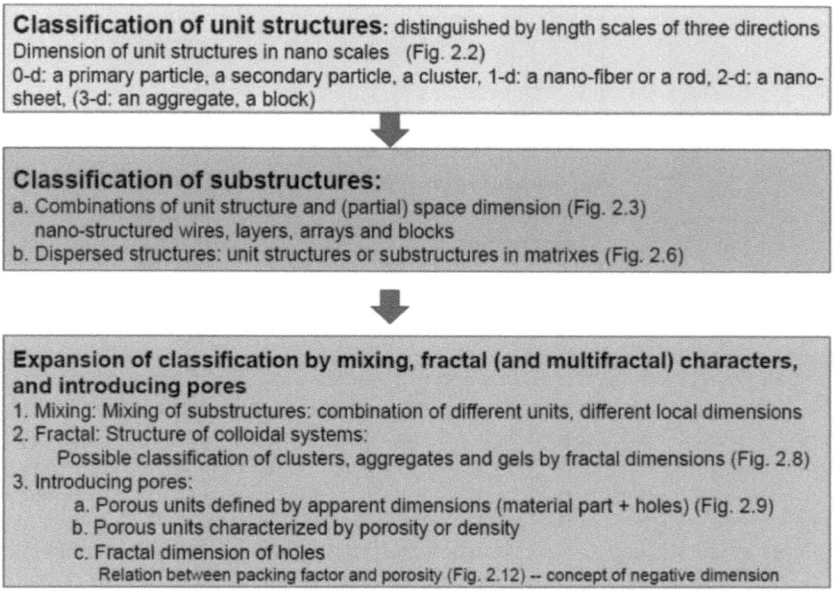


**Figure 2.1** Outlines of classification and expansion of it to fractal (multifractal) and porous systems.

In Fig. 2.2, elemental units of the structures with different dimensionality are shown. They are named as 0-d~3-d with using small character d.

As in ref. [7], if all length scales, $l_x$, $l_y$, and $l_z$ are less than or equals to $\lambda$, the unit can be regarded as zero-dimensional (0-d). The unit of 0-d in Fig. 2.2 can be colloids, primary-particles, clusters. Combination of 0-d structures can be still in the nanoscale. Therefore, secondary particles in nano-size can be still regarded as 0-d structures. The 0-d units can form structures of different dimensionality as shown in Fig. 2.2. If two directions are less than $\lambda$, the unit is regarded as 1-d, where structures are formed by combinations of 0-d units. If one direction (thickness) is less than $\lambda$, it is regarded as 2-d, where the unit is formed by 0-d or 1-d units. 3-d units are not in the nanoscale but can be a part of the substructures in nanostructured materials.

This concept is similar to those in refs [7, 8] in the sense that it is based on the dimensionality of the structural units.

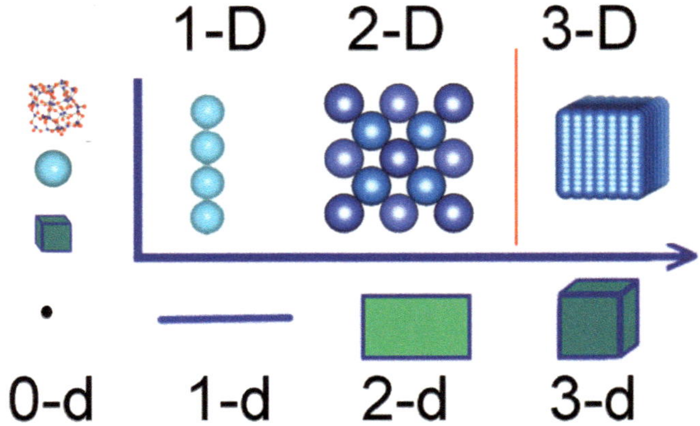


**Figure 2.2** Schematic descriptions of elemental units of nanostructured materials, where the dimensionality is determined by its length scale of different directions. [0-d structures]: These elemental units are regarded as zero dimensional (0-d), which is atomic, or clusters or molecular levels. The elemental units can be primary particles, colloids, if $l_x$ $l_y$ $l_z$ are less than or equals to $\lambda$. [1-d structures]: If a diameter of the unit is less than or equals to $\lambda$, it can be regarded as a nanostructure unit (a nanofiber). The unit is regarded as 1-d, where structures are formed by combinations of 0-d units. Several length scales can be used to distinguish, for example, short and long structures. Here the value $\lambda$ is a length scale determining the range of sizes to treat as nanomaterials. [2-d structures]: If one direction (thickness) or a unit is less than or equals to $\lambda$, it is regarded as 2-d, where the unit is formed by 0-d or 1-d units. Several sized ones can be distinguished if necessary. [3-d structures]. These structures are not in nano-size but can be a part of nanostructured materials.

### 2.2.2 Substructures of Nanostructured Materials

Substructures formed by these elemental units in each space dimension, D, are considered. Some examples for substructures (images) by this classification using dimension of structural units and that for space are shown in Fig. 2.3 as a combination of rows and columns. The essence of descriptions of elemental units in Fig. 2.2 is depicted in the bottom row of Fig. 2.3. Different from Fig. 2.2, substructures in Fig. 2.3 contain assembly and/or hierarchy structures of elemental units. All substructures shown in Fig. 2.3 can be regarded as building units of NSM. All possible combinations can be considered based on this kind of figure (including porous systems mentioned later).

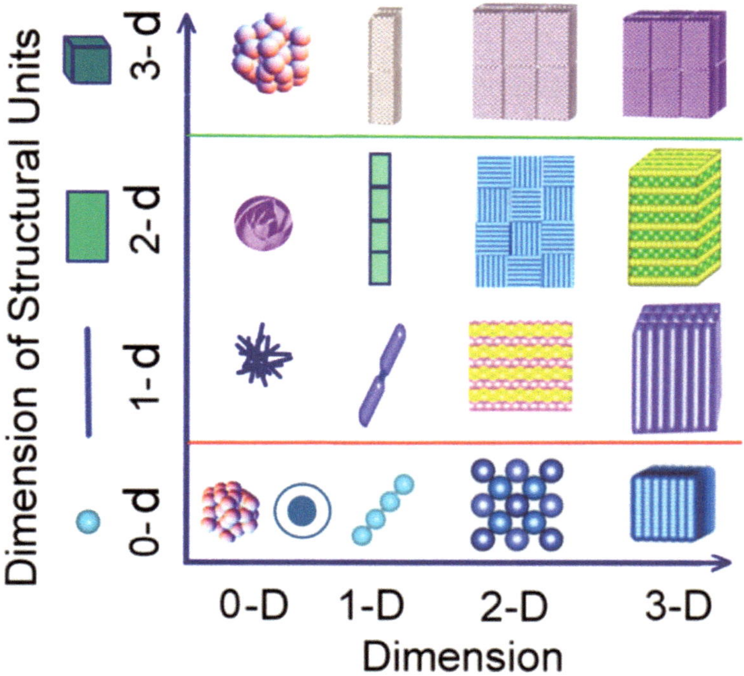


**Figure 2.3** Substructures of NSM: Schematic description of classification of substructures of nanostructured materials based on the dimension of structural units, (0-d~3-d) and that of the (partial) space occupied by the substructure. Structures thus defined can be indexed by both dimensions. For example, arrays of columns forming a 3D structure can be indexed as 3D(1d). The difference in colors within each substructure is for a distinction of different chemicals and/or properties but not necessarily mean different constituents. If 1-d structures (sticks, rods), are in 3-D space and formed 3-D structures, one can use the index like 3D(1d). Elemental nanostructural units and their assemblies are included in the bottom row (below a red line). The combination of 0-d and 0-D can represent secondary particles. The core–shell-type particle can be also included in the class of 0D(0d). The rows for 3-d (above a green line) are not in nano-size but can be a part of the nanostructured materials. 3D(0d~2d) are not necessarily in nano-size but they can include nanostructured parts. Expansion of the classification, including mixing of these structures, fractal dimensions of assembly of structures, and contribution of pores, will be considered later.

Substructures thus defined can be indexed by both dimensions. If 1-d structures (sticks, rods), are located in 3-D space and formed 3-D structures, one can use the index like 3D(1d). Merit to use such indexing is not only to classify already-existing materials but also to find out missing combinations of building units with different characteristics. In these figures, space dimensions 0-D, 1-D and 2-D are for a partial space of 3-D.

## 2.2.3 Substructures with Different Dimensionality

### 2.2.3.1 Treatment of nanoparticles, nanofibers, nanosheets, and nanoflowers in the classification

Nanoparticles are regarded as quasi 0-dimensional materials as in historical treatment. A combination of 0-d structures may form larger sized nanoparticles or clusters. They can be still in nano-size, if side lengths $l_x$ $l_y$, and $l_z$ are less than $\lambda$ or equals to $\lambda$.

In some of the systems, the size of the particle is negligible for the properties determined by $\lambda$. However, caution is required for the fact that the volume or surface of such particles cannot be neglected in many cases. As in a similar manner, the size of 1-d (fibers, rods) and thickness of 2-d sheets (layers) should be treated carefully.

Additional explanations may be required on how these units are combined.

In Fig. 2.3, the core–shell-type structure can be included in the 0D(0d). This kind of structure is treated as 0D(00) in ref. [7] and is treated as 0D material in ref. [8]. Secondary particles, which are still in a nano-size, can be also represented like this.

"Nanoflowers" can be included in the class 0D(2d). Many interesting structures (images obtained by scanning electron microscope (SEM) or transmission electron microscopy (TEM)) including nanoflowers are found in ref. [19].

In the following, the classification will be expanded to include mixtures and fractals. Further treatment of heterogeneity will be considered with multifractal formalism.

## 2.2.4 Mixing or Coexistence of Substructures

### 2.2.4.1 Simple mixing of substructures with the same dimension

When more than one substructure (in Fig. 2.3) is included, it is indexed by combinations of these substructures. We assume that the contents and size in these substructures are comparable at least in order. In such cases, repeating indexes will be good enough and there are no limitations for the number of constituents in our treatment. The system of 2-D sheet contained nanoparticles are indexed as 2D(0d,2d). It can be represented as 2D(2d)+2D(0d), if one prefers to emphasize that it is a mixture of two substructures.

In the figure, colors are used just to distinguish the different parts of units and not mean heterostructures. However, of course, the combinations of different constituents are possible. Additional information such as thickness, the order of laminations will be added if necessary. For example, the 2D-layers formed by two different constituents (a and b) can be represented as $2D(2d_a2d_b)$ or simply $2D(2d_{ab})$, while the NSM of 2D-layers with six components can be represented as $2D(2d_{abcdef})$.

Almost of 36 classes except for skeleton and honeycomb type (they will be included in porous systems later) structures found in ref. [7] can be prepared by combinations of structures in Fig. 2.3.

As another example, a mixture of 3D(1d), 3D(0d), and 3D(3d) is considered here. The index of the mixture, 3D(0d,1d,3d) or 3D(0d1d3d), can be easily compared with No. 34 (3D310) defined in ref. [7].

#### 2.2.4.2 Combinations of substructures with different dimensions

Any combinations of dimensionality of substructures are allowed in this treatment.

For example, stacked pipes (3D(1d)) can be wrapped by a sheet indexed by 2D(2d). Such a situation can be represented as 3D(1d)+2D(2d). To consider possible combinations will bring new developments of NSM.

#### 2.2.4.3 Treatment of the case with the lower dimension of the substructures than elemental structural units

In ref. [7], constituents having larger dimensionality than the space dimension were excluded. It is a natural choice; however, such cases are not excluded *a priori* here, because new designs of nanostructured materials might be found there. How can higher dimensional units form quasi lower dimensional structures? Just one example is given here. Tetrahedron of $SiO_4$ units forms networks with several dimensionalities (see Fig. 2.4 as an example of chain structure) as found in silicates in both crystals and glasses [20, 21]. Sometimes, $SiO_4$ units in silica, silicate, and many composites seem to be better to be treated as tetrahedral-shaped units. In the case of $SiO_4$ units, it can be treated as 0-d as well as 3-d by a choice of $\lambda$. Other new 3-d units, which can form lower-dimensional materials, can be designed.

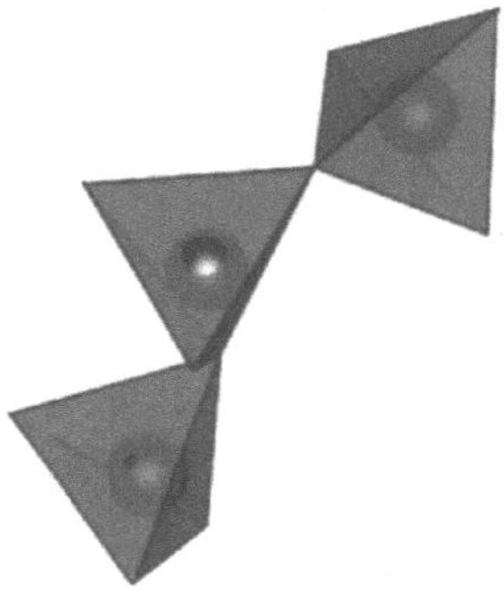

**Figure 2.4** An example of the structure, where the 3d unit forms a quasi-one-dimensional structure. Figure is taken from the structure found in MD simulations of silicate glass. In this case, tetrahedral units of $SiO_4$ are connected by corners. Similar situations, where dimensionality of nanostructural units is larger than the (partial) space dimension, are allowed in Fig. 2.3.

#### 2.2.4.4 Combination of elemental nano units with different dimensions

In our treatment, the combination of 0-d~3-d units with different dimensionality is allowed for mixing. The merit of this inclusion can be found in the following example.

The instantaneous structure of ionic liquids, 1-ethyl-3-methyl imidazolium nitrate (EMIM-$NO_3$), is shown in Fig. 2.5, which is obtained by MD simulations [1, 22]. A substructure of a cation can be an imidazolium ring (2d) and chains (methyl and ethyl) attached it. If we neglect some details, chain part can be indexed (+1d). $NO_3^-$ ions are regarded as 2d. Then the possible index of structural units can be 3D((2d+1d),2d). One can modify it for different purposes following the use. For example, other properties such as sizes, colors, charges, length of chains, the magnitude of the disorder can be added to characterize each component.

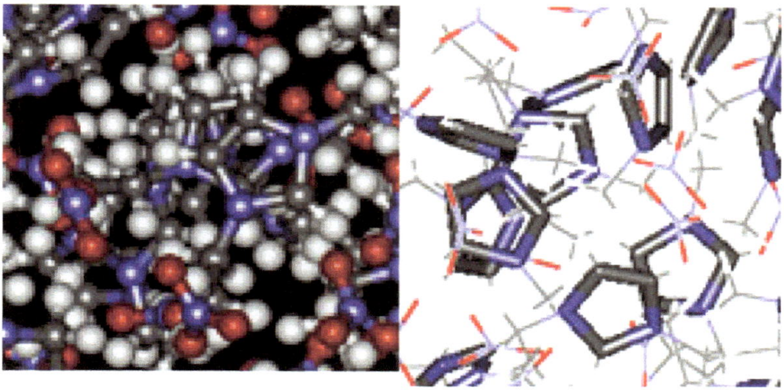

**Figure 2.5** An example of combination of structural units. Structure of ionic liquids, 1-ethyl-3-methyl imidazolium nitrate (EMIM-$NO_3$), with ball and stick representation (left) and wireframe representation (right) both obtained by MD simulations [1] are shown. The structural unit can be a combination of imidazolium ring (2-d) and chains (methyl and ethyl) combined to the ring. If we neglect some details, the chain part can be indexed (+1d). A structural unit of $NO_3^-$ ions is regarded as 2-d. Then one of the possible indexes of structural units can be 3D((2d+1d),2d).

### 2.2.5 Dispersion of Nanoparticles in Matrices

Let's consider the nanoparticles on/in the 2d sheet. It can be regarded as a mixture of 0-d units in 2D and possible representation of index is 2D(0d,2d) as already mentioned. However, if the nanoscale units are much smaller in amount than other components and spread in the system, it may be better to regard as dispersoids in matrixes (sub-index m is used here). Since similar situations are often encountered in nanotechnology as in colloidal solutions, it may be useful to treat them separately, if necessary. As shown in Fig. 2.6, 0-d~2-d units can be dispersed in matrixes, where the matrix can be any structures in each column (1-D~3-D) in Fig. 2.3. Since 3-d structural units are not in nano-size and hence are not treated as the dispersoid in Fig. 2.6. Still, they can be a part of structures.

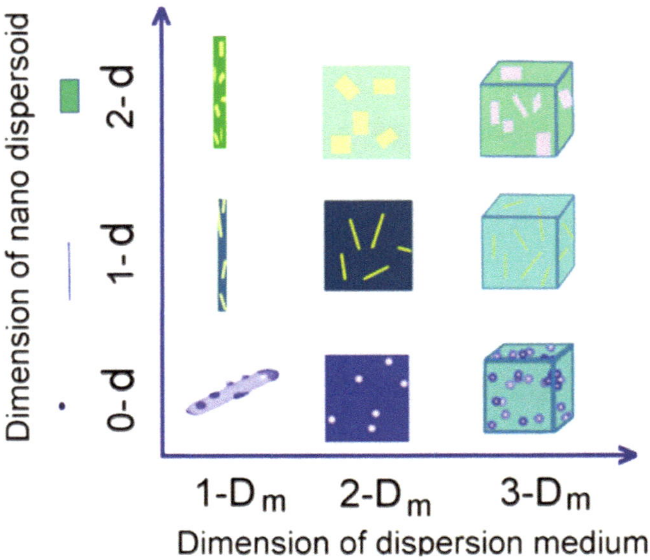


**Figure 2.6** If the content (and size) of elemental nano units is much smaller than another constituent, it may be regarded as dispersoid. Schematic description of dispersion of nano units in the dispersion medium with different dimensions is shown here. Dispersion medium can be any of substructures on the corresponding column in Fig. 2.3.

In a similar manner, 1-d and 2-d units can be dispersed in matrixes. Since 0-d and 3-d structures are different just in sizes, situations of the 3-d are comparable except for sizes and therefore larger 3-d structures than nano-size were excluded from the dispersoid in this figure.

Since 1-d structures are defined by the radius, the existence of both long fibers and short (in nanoscale) fibers is allowed. It is a possible choice to divide them into two (or more) subclasses by its length, if necessary. A similar argument holds for 2-d for two directions of longer length sides. In the case of 0-d, secondary particles, clusters (0D(0d) structures), and other substructures (except for 0~3D(3d)) in Fig. 2.3 can be dispersed in matrixes. In the case of nanoparticles on the nanosheets, it can be a mixture of two components as already mentioned. If one would like to emphasize that 0-d structures are a minor component as in the case shown in Fig. 2.6, where the matrix is of type 2D(2d), the expression can be (0d),[$2D_m$(2d)] or 0d[$2D_m$(2d)].

The case for 0-$D_m$ is not included in the figure; however, for some purposes, it can be included.

To simplify the expression, the addition of sub-index for dispersoid ($_d$) may be useful. In the case of the 0-d structure in 2D(2d) mentioned above, it can be represented as 2D(2d,0$d_d$).

#### 2.2.5.1 Treatment of network systems

Although the $SiO_4$ units in silica and/or silicates can be represented as the elemental units 0-d or 3-d (by choosing suitable $\lambda$) units as already mentioned, further consideration seems to be necessary for inclusions of network systems.

With increasing alkali metal contents in the composition of $xM_2O$-$(1-x)SiO_2$-type systems (M means alkali metal), the dimensionality of network structures is known to shift gradually, in the order, 3-dimensional networks, 2-dimensional sheets, 1-dimensional chains formed by the connection of $SiO_4$ units and quasi 0-d (separated $SiO_y$) units. Corresponding (but in inversed order) shift occurs for connection of $MO_z$ (M = Li, Na, K–) structures. Similar situations are found for other network-forming materials such as borates, phosphates, and so on, in the molten, crystalline, and glassy states. Many nanostructured materials contain several network structures, and therefore the dimensions or fractal dimensions of network structures are important measures of the complexity of the structures. That is, for the treatment of network-forming materials, further expansion of the classification including fractals will be useful.

Structures of $SiO_4$ networks are also governed by the packing in space. It is well known that the systems connected by faces are more packed than the connected by edges, while units connected by edges are more packed than the system with sharing a corner [20, 21]. Therefore, we must use some indexes to distinguish the packing in the structure of network systems and/or other systems. Thus, the structures considered in Fig. 2.3 are assumed to be dense (well packed) ones and later porous systems will be treated separately, even though it can include the case such as carbon nanotubes, where a 2-d sheet forms a 3-D structure as 3D(2d) if the role of the empty part is not focused.

For such network structures, $Q_n$ (where $n$ means the number of bridging oxygen atoms in $SiO_4$ units) distributions [22–25] are useful for characterization of the system (see Chapters 4 and 9.)

Besides the expansion of the classification to fractal structures, expansion to porous structures with fractal natures and differences in packing is a possible choice.

Nowadays, various systems have been devised besides traditional network systems. Especially, several kinds of metal organic frameworks (MOF) [26, 27] (metal ions or clusters coordinated to organic ligands) with several dimensionalities are designed and they can be regarded as elemental building blocks of complicated structures. They are also building blocks of porous materials [26–28] with large surface areas.

## 2.3 Expansion of the Classification Using Fractal Dimension for Assembly of Structural Units

Many NSMs might have more complex structures and not necessarily in the crystalline state. In Chapters 5–9, the following nanostructured materials are treated.

1. Quasi-equilibrated systems such as lithium silicate glasses and those after introducing pores, which have complicated structures consisting of ions surrounded by oxygen atoms and network of silica combined with them
2. Nanocolloidal systems in the aqueous solution of salt and nanostructured system after aggregation and gelation having several partial structures

Porous systems treated in Chapters 6 and 7 in case 1 mentioned above are starting from glassy systems and have both disorder and porosity, while systems such as gel and aggregate (Chapters 8 and 9) in case 2 show fractal shapes. Now a target of debates is disordered systems and not polycrystalline nor partially crystalized systems, although methods and concepts discussed in the book apply to both.

In these systems, changes among related systems are rather gradual and may have complicated heterogeneous structures. In such cases, the dimensionality of assembly of structural units will be discussed by the fractal dimension [4, 5] rather than dimension of space. In the following sections, classifications will be expanded to fractal (and multifractal) and porous systems.

## 2.3.1 Classification by Using Fractal Dimensions

### 2.3.1.1 Structures of spontaneously formed gel obtained by full atomistic MD simulations

In Fig. 2.7, an example of the structure of the gel obtained from the full atomistic MD simulations of silica-nanocolloid–water–NaCl systems [4] is shown for a slice. As shown in this figure, several components and networks formed complicated structures. Some characteristics of structures are picked up here. First, the silica part forms infinitive networks (with periodic boundary conditions of MD). Na and Cl partially form networks but are surrounded by networks of water. The water part and the silica part also interact. That is, the silica network has a complicated shape and the water part is surrounded by it. In Chapters 8 and 9, the structure will be characterized by several methods.

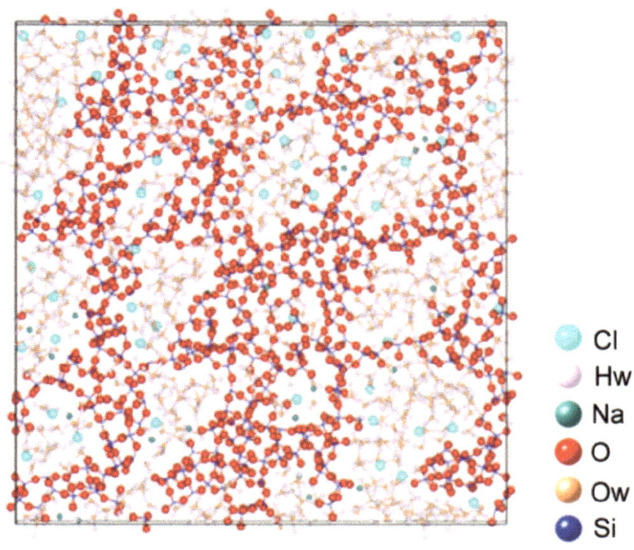


**Figure 2.7** An example of a nanostructure obtained by Molecular Dynamics Simulations. A slice (with a thickness of 10 Å) of the structure of spontaneously formed gel from silica-nanocolloid–water–NaCl systems [4]. Details for this simulation and obtained structures will be shown in Chapters 8 and 9. Here the Hw and Ow mean H and O atoms in water.

At first glance, the water part seems to be separated by the networks formed by $SiO_4$ units in a gel, when it is shown by the slice; however, the water part is connected to the next slices, if we examine the structure in detail. As shown in this example, 3-D connections of structures may be missing in the representations, by a slice or by projection to a plane. Therefore, as it will be discussed in Chapter 9, caution is required to treat fractal structures by 2-D images (such as transmission electron microscopy (TEM) images).

How can the structure of the gel be classified? One of the possible choices for expansion of the classification will be achieved by introducing fractal dimension [9, 10], including the concept of multifractal [11–13] for assembly of structural units. Concepts of fractal and/or multifractal and methods of fractal dimension analyses will be shown in Chapter 4.

#### 2.3.1.2 Monofractal and multifractal structures and classifications

The units to form a fractal structure can be any of the ones in Fig. 2.3. The systems with matrixes in Fig. 2.6 can be also a target of such treatment.

If the structure is monofractal, it is simply characterized as one fractal dimension, $d_f$. If the 1-d structure formed the fractal structure with the fractal dimension 2.5, then the index such as $2.5d_f$(1d) can be used.

In the case of gel in [4], fractal dimension of the silica part for a short length scale was found to be 2.25, while that for the longer length scale was found to be 3.01, when we used the running coordination numbers of Si–Si pair to calculate them (see Chapters 8 and 9). Existence of more than one exponent means the multifractality of the system (details of the explanation of this concept will be given in Chapter 4).

In Fig. 2.8, the $d_{f\text{-long}}$ values are plotted against $d_{f\text{-short}}$ values for networks of the silica part of systems obtained by atomistic simulation of silica-nanocolloid–water–NaCl systems with several conditions. Systems that include clusters, aggregates, and gels are found in the silica-nanocolloid–water–NaCl systems of series I and II. Here, contents of salt change in the order of $A<B<C$ in

each series. Series II systems are more water-rich than series I. II-A is a reference system without salt. In II-A, several clusters are found, while aggregates are developed with the concentration of salt as shown in II-B and II-C. A large aggregate developed in I-A. When the system I-B was annealed at 400 K, gel was formed in I-C.

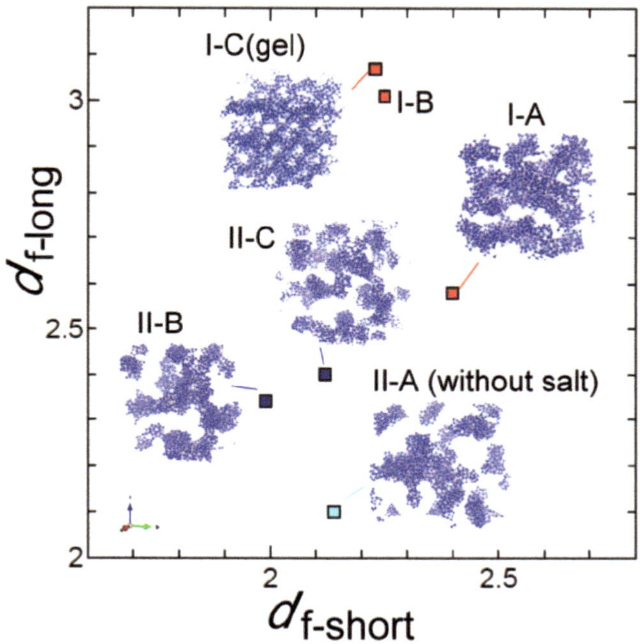


**Figure 2.8** Structures of silica part for clusters, aggregates, and gel obtained from molecular dynamics simulations of silica-nanocolloid–water–NaCl systems (see Chapters 8, and 9 for details). Fractal dimensions of the longer length scale are plotted against fractal dimensions of short length scale regions for these structures. Then these systems can be characterized by the combinations of these fractal dimensions. Series II (Blue squares) are more water rich than series I (Red squares). II-A (A pale blue square) is a reference system without salt. Contents of salt are in the order of A<B (before gelation) =C (after gelation) in series I, while they are in the order of A<B<C in series II. Coagulation developed in the order of II-A<II-B<II-C<I-A<I-B<I-C. Methods of analyses of fractal dimensions are shown in Chapters 4 and 9. One can understand the development of the coagulation process by the position on this plot. Therefore, the plot can be used for expanded classification of complex systems.

The plot can be used for classification by the multifractal character of the system (coexistence of different fractal dimensions). That is, the situation of the local dense structure can be distinguished from the global connection of such structures by their fractal dimensions. Therefore, to classify networks of the gel, one of the possible methods might be by the combination of two exponents, 2.3 $d_{\text{f-short}}3.0d_{\text{f-long}}$(0d), (with dimensions of short and longer regions and dimension of the unit regarded as a 0-d).

When $d_{\text{f-long}}$ value increases and exceeds a certain value, the gel will be formed. The value seems to be ~2.5, which is comparable to the threshold value for the 3-D percolation (2.53 [29]), or a related one. In this example, systems can be classified for colloids, clusters, aggregates, and gels. That is, the value of $d_{\text{f-long}}$ changes in the order of cluster<aggregate<gel; and connectivity of long-ranged structures is well represented by $d_{\text{f-long}}$, while the values for $d_{\text{f-short}}$ seem to show fluctuation related to the regularity of the porous parts of the pattern (lacunarity) [30].

Multifractality can be regarded as a kind of the presentation of the heterogeneity [1], and hence, the plot using both length scales can classify heterogeneity of systems. Generally, some characteristic regions can be defined on the plot, if there are enough number of data points. By using plots, one can classify each system by its region where it belongs to (see Chapter 9). Thus, mapping of related and other systems will be useful for classifications of complex systems as well as for clarifications of the mechanism of coagulation. Another possible way of classification can be based on multifractal analyses [14–16]. Parameters of singularity spectra, $f(\alpha)$, can be used for such purpose. Examples of analyses as well as applications to other systems will be discussed in Chapter 4 for the case of porous systems.

### 2.3.2 Characteristics of Porous Materials

In ref. [7], inverse nanostructures with cavity building units are not regarded as a separate one. It seems to be a reasonable choice to simplify the classification. However, porous materials are an important class of materials in many kinds of applications, and properties of such systems are different from the packed ones.

Therefore, we consider it separately and then relations among these cases are discussed later.

Structures of pores are closely related to the mechanical properties, electrical properties, solubility, and characteristics for storages and can be characterized by their shapes, connectivity, porosity, and lacunarity [30] and/or fractal dimensions. Porous materials can be subdivided into three categories by sizes of pores, set out by IUPAC [31]. That is, the pores with widths exceeding about 50 nm are called macropores; the pores of widths between 2 and 50 nm are called mesopores; and the pores with widths not exceeding about 2 nm are called micropores.

Also, in the recommendation by IUPAC [32], authors can classify the pores according to their availability to an external fluid (closed pore, open pores, blind pores) and/or shapes (cylindrical, ink-bottle shaped). In the report, the possibility to use fractal dimension analysis for the characterization of porous systems is also referred to. That is, characteristics of surface properties such as pore volume are considered to be proportional to (resolution of analysis)$^D$, where $D$ is the fractal dimension of the surface for which the property is relevant.

Porous materials gain the attention of researchers in many fields. For example, transport in the porous media is discussed by Adler and Thovert [33] and the importance of the fractal characters is discussed. The character seems to be necessary to clarify the effects of nano-sizing on several properties. O'Neill and coworkers [34] have examined the porous materials as a candidate for low-dielectric-constant materials, which is useful for IC manufacturing at 45 nm features and beyond. In that work, MD simulations (in *NPT* conditions) were used to examine the relation between pore size and mechanical properties. In their simulations, the modulus decreases with porosity as $\sim(1 - \varphi)^5$, where $\varphi$ is the porosity of the system. This relation is more rapid in most oxides with macro porosity and faster than the upper bound of the Hashin–Shtrikman theory [35, 36] as discussed by them. Hence, the classification of porous materials will be useful for systematic studies of the properties of them. Furthermore, the interconnectivity of pores is an important property in many applications of porous materials. In the percolation theory [29], the percolation threshold for two-phase systems is known to be ~0.16, if we use the packing fraction (in volume) as an index. Thus, the connectivity is closely related to the porosity.

### 2.3.3 Classification of Porous Materials Based on the Apparent Dimension of the Structures

In each class of structure, systems with different porosity may exist or can be prepared. In this section, the classification of porous systems will be tried similarly as in Fig. 2.3.

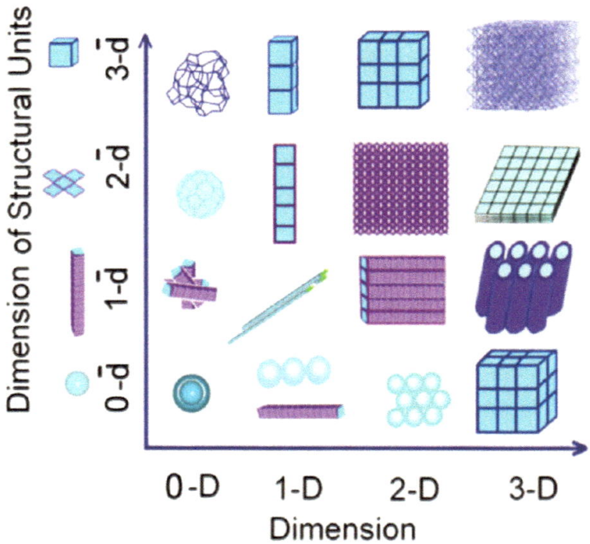


**Figure 2.9** Examples of substructures of porous materials: Classification is by apparent dimension of the structures plotted against space dimension. Possible measures to characterize structures are porosity in each class, fractal dimensions and so on. The index can be like 3D(1d̄), for the example of the stacked pipes (The third case in the right column.). As shown in the next figure, dimension of the porous materials can be measured for both material and holes. Characteristics of apparent structures are analogous to the structure in Fig. 2.3 except for pores and therefore, the explanations in Fig. 2.3 for each position is applicable. These structures are comparable to the situations in Fig. 2.3. In a similar manner, dispersions of holes can be represented as in Fig. 2.6.

In Fig. 2.9, examples of typical structures of porous materials are shown. In this figure, the classification is done now by apparent dimension (formed by both rigid and porous parts) of the structures plotted against the space dimension. The index can be determined like 3D(1$\bar{\mathrm{d}}$). Here $\bar{\mathrm{d}}$ is the apparent dimension of the structure containing the porous part. As shown in Fig. 2.10, the merit of this indexing is the simplicity of the expression of the combination of dense and porous materials with distinguishing characteristics of each component. An example on the left-hand side of the figure is for stacked pipes with and without holes, which can be indexed as 3D(1d,1$\bar{\mathrm{d}}$). On the right-hand side of the figure, the structure corresponds to a kind of core–shell particle [11] with pore and rigid parts and can be indexed as 0D(0d,0$\bar{\mathrm{d}}$). One can change the index order by size or by other properties.

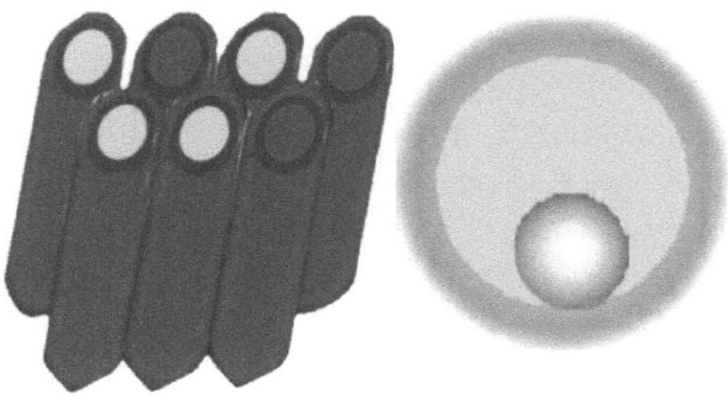

**Figure 2.10** Examples (images) of possible combinations of porous and non-porous substructures; left: coexistence of rods and pipes, where only diameters are in nano-size; right: a nanoparticle in a nanopore. Using the classification of porous systems as in Figs. 2.3 and 2.9, both cases can be treated, qualitatively. The possible indexes for left and right cases are 3D(1d,1$\bar{\mathrm{d}}$) and 0D(0d,0$\bar{\mathrm{d}}$), respectively. Other possible methods for indexing of porous materials are discussed in the text.

These situations are not the same as classification in ref. [7]. Both skeletons of fibers and nanotubes are classified into 3D1 in that reference based on the dimensionality of the material part surrounding the voids, while in our classification, the skeleton of fibers is indexed as 3D(3$\bar{\mathrm{d}}$) while nanotubes are indexed as 3D(1$\bar{\mathrm{d}}$), respectively. This distinction between the two cases seems to be useful to consider mechanical and related properties.

Of course, one of the possible measures to characterize porous structures is density and/or porosity in each class as used in many works. For more quantitative treatment, the index such as 3D(1d($\varphi = a$)) can represent stacked pipes, where $\varphi$ is the porosity.

In Chapters 6 and 7, porous silicate systems are treated simply as a function of density, since the density was found to be a function of porosity. In the case with porosity, the change of the shape of the system with introducing pores is not directly included. Naturally, the classification can include combinations of materials with different porosity.

## 2.4 Relation between Fractal Dimension of Holes and Porosity

Next, we will consider the classification including porous materials from a different point of view. When the shapes of pores change with porosity, fractal shapes might be found. Previously, the fractal dimension of the silica part in the gel was used for the classification of materials. A similar treatment is possible for the water part as well as the salt part, if necessary. If one removes the water part from the structures of colloidal solutions after gelation, a porous structure is obtained. Thus, the structure of the pores will be comparable to that of the water part. A similar argument holds for ion channels in ionically conducting glasses. For instance, as shown in Fig. 2.11, the shape covered by the trajectories of Li ions in the $Li_2SiO_3$ glass is comparable to the ion channel formed by networks of $SiO_4$ units. Thus, it is quite natural that the structures of holes correspond to those of filled materials within the holes. This means the fractal dimension analysis for pores, cavity, and holes is applicable reasonably. Now the definition of the dimension is expanded to include fractal dimensions of assembly of structural units.

In general, a packing factor can be defined as the fraction of materials in a container with a total volume of $V_{\text{total}}$, that is $V_{\text{m}}/V_{\text{total}}$. For example, the atomic packing factor in the crystal structure is the fraction of volume in a unit cell. On the other hand, porosity is usually defined as a fraction of pores or voids in the material as already mentioned, that is, it can be defined as $V_{\text{pore}}/V_{\text{total}}$. Thus, porosity is a counterpart of the packing factor. In this case, we can define two kinds of fractal dimensions even in a simple system. That is, one is for the structure of material and the other is the structure of holes (pores, channels).

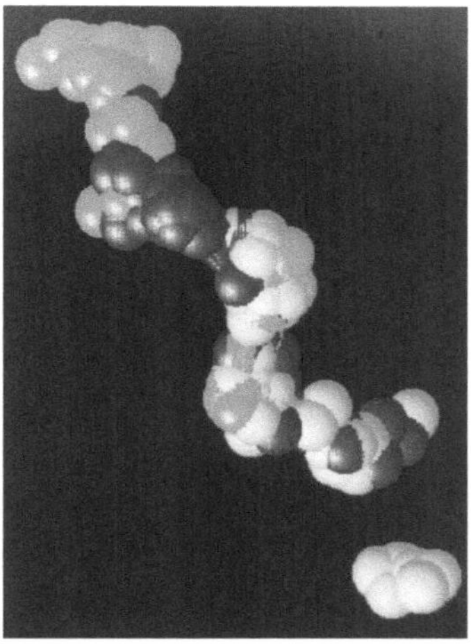

**Figure 2.11** Structure of ion channel visualized by accumulated positions of three Li ions in lithium metasilicate glass obtained by MD simulations during 1 ns run at 700 K. Each ion has a different color. The shapes of the paths of ions correspond to the channels formed by the network of $SiO_4$ units. Zig-zag shapes of ion channel are clearly found by the trajectories of ions. As in a similar manner, holes in the gel can be visualized as the water part filled in gels (see Chapter 9).

In Fig. 2.12, the situation of different packing factors (or fractions) is schematically compared with different porosities in the system. As found in Fig. 2.12a, if the content of material increases, the packing factor increases and fractal dimension of the material will also increase. One can also consider the situation with different contents of pores. An opposite but a quite similar situation can be seen for the increase of pores, cavities, channels (hereafter we use the word "holes" to include these different shapes) in the system as shown in Fig. 2.12b. With increase of the porosity of the system, the content of the holes increases. With this change, fractal dimension of holes will increase. Yu and Li [37] have examined the fractal character of the porous media, and it was shown that the fractal dimension $D_f$ is a function of porosity, $\varphi$, and the ratio of the lower limit to the upper limit of self-similar regions for Sierpinski carpets. That is,

$$D_f = 2 - \frac{\ln \varphi}{\ln \dfrac{\lambda_{min}}{\lambda_{max}}}, \tag{2.1}$$

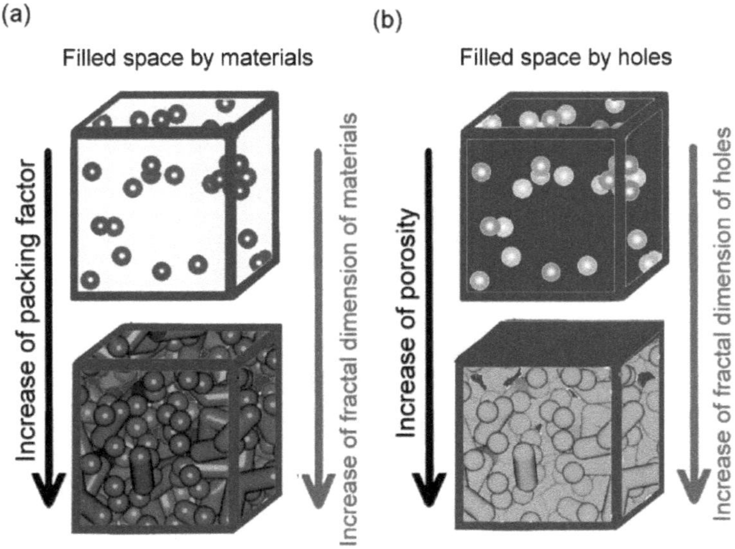


**Figure 2.12** Concept of negative dimension: Relation between the packing factor and porosity in 3D space. (a) In a real space, with an increase of packing factor of contents, particles tend to form larger units, clusters. With this change, the fractal dimension of materials also increases. (b) Opposite but a quite similar situation can be seen for the increase of holes (pores, cavities, channels) in the system. With increase of porosity, fractal dimension of the empty part increases. If one considered the filled cubic as the standard point, increases of fractal dimension of holes can be regarded as the increase of negative dimension. Another possible method for the classification of porous materials is thus by "fractal dimension of holes" or by using "negative dimension." Note that this concept is different from that used in Fig. 2.9.

and for Sierpinski gaskets,

$$D_{\mathrm{f}} = 3 - \frac{\ln\varphi}{\ln\dfrac{\lambda_{\min}}{\lambda_{\max}}}, \tag{2.2}$$

where $2 < D_{\mathrm{f}} < 3$. They argued that these relations hold for a disordered system having statistically fractal nature. Since the porosity is the counterpart of the packing factor, comparable equations should hold for the relation between fractal dimension of material and the packing fraction of it.

### 2.4.1 Concept of the Negative Dimension

If one uses "fractal dimension of holes" instead of "fractal dimension of the structure formed by substances," the dimension might be regarded as the negative dimension [38]. In other words, if we considered the filled cubic by a matrix as the standard point in Fig. 2.12, "fractal dimension of holes" can be regarded as the negative dimension. Let's consider the situation of carbon nanotubes mentioned before. It can be classified by the dimension of the hole, if one would like to emphasize the porous part of the structures and it might be indexed as 3D(-md) or $-md_f$. Here "–m" means the fractal dimension of holes.

## 2.5 Different Kinds of Classifications

These classifications and definitions of units suggested in Sections 2.2 and 2.3 seem to be general but still are not definitive.

For example, in the case of magnetic properties, Leslie-Pelecky and Rieke [39] have proposed a classification shown in Fig. 2.13. Here the classification was designed to emphasize the physical mechanisms responsible for the magnetic behavior.

Properties of systems can also be treated as functions of other characteristics of the systems.

Core–shell particles seem to be an important class of NSM, which consists of two or more materials with the core part and surrounding part. In ref. [11], core–shell-type particles are classified by using different criteria. One method is based on the combinations of organic and inorganic materials for constituents. That is, four classes, (i) inorganic/inorganic, (ii) inorganic/organic, (iii) organic/inorganic, and (iv) organic/organic can be defined.

Another classification found in that reference of core/shell nanoparticles is as follows:

(a) spherical core/shell nanoparticles; (b) hexagonal core/shell nanoparticles; (c) multiple small-core materials coated by single-shell material; (d) nanomatryoshka material; (e) movable core within hollow shell material.

Within the second classification, (a)–(c) can be indexed by our scheme, with the addition of subindexes for shapes and/or numbers of elements. Interesting structure found there is (d) nanomatryushka material. The hierarchy of structures can be represented by the hierarchy of parentheses in our indexing like 0D(0d(0d(0d))). The structure (e) seems to correspond to the right-hand side of Fig. 2.10.

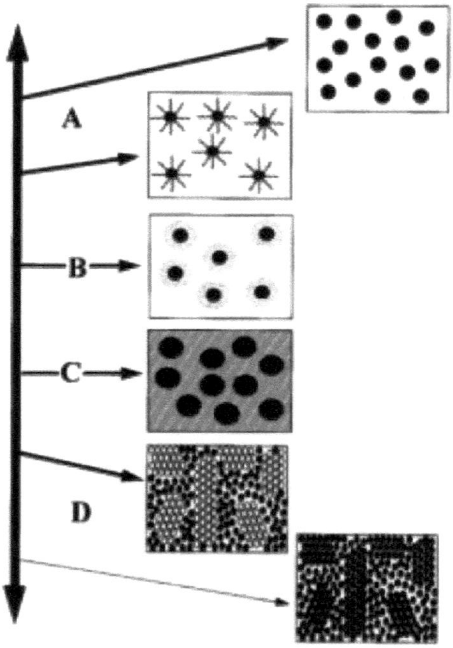


**Figure 2.13** Schematic representation of the different types of magnetic nanostructures. Type-A materials include the ideal ultrafine particle system, with interparticle spacing large enough to approximate the particles as noninteracting. Ferrofluids, in which magnetic particles are surrounded by a surfactant preventing interactions, are a subgroup of Type-A materials. Type-B materials are ultrafine particles with a core–shell morphology. Type-C nanocomposites are composed of small magnetic particles embedded in a chemically dissimilar matrix. The matrix may or may not be magnetoactive. Type-D materials consist of small crystallites dispersed in a nanocrystalline matrix. The nanostructure may be two-phase, in which nanocrystallites are a distinct phase from the matrix, or the ideal case, in which both the crystallites and the matrix are made of the same material. Reprinted with permission from Leslie-Pelecky, D. L., Rieke, R. D. (1996), Magnetic properties of nanostructured materials, *Chem. Mater.*, 8, pp. 1770–1783. Copyright (1996) American Chemical Society.

## 2.6 Common Possible Measures of Nanostructured Materials

### 2.6.1 Effects of Nano-Sizing

Some fundamental effects of the nano-sizing considered in this chapter are as follows:

1. Size effects: There are several kinds of size effects so far argued.

   In ref. [6], Gleiter argued that size effect is observed when the size of the system is comparable with the characteristic lengths of the phenomena. Here characteristic lengths may depend on the property to be discussed. Namely, it can be de Broglie wave (in the case of quantum dots), correlation length, length among sites, *etc.*

   The size effects of nanoparticles tend to be argued with extremely large surface areas. Some of the size effects related to the surface areas/regions will be argued later.

   Besides such effects, finite-size effects are known for electrostatic calculation as well as molecular dynamics simulations. Finite system size effects in molecular dynamics simulations are caused by the periodicity of MD cells, imposed by periodic boundary conditions.

   One should not forget that if the system size is reduced, the length scale found in the system is cut or modified. That is, the structure itself is modified by the system size. Therefore, a comparison of system size with the characteristic length of the phenomena is not necessarily enough to discuss the size effects. Some methods to measure length scales from MD data will be introduced later.

2. Change of the dimensionality of the system: The change of dimension in a nanosystem is concerned with the dimension of the structural units as well as the formation of larger structures such as clusters, aggregates, and gels. Therefore, this concept includes fractals and multifractals [14–16]. That is, the complexity of the structures containing more

than one exponent is considered. Examples for gel and related systems will be shown. Several methods for measuring fractal dimensions for both the structure and dynamics of systems based on the MD data will be explained in Chapter 4. The dimensionality of the system has been already used as the basis of the classification mentioned in the previous sections because the properties of nanostructured materials are affected by their dimensionality.

3. Atomic structure near the surface: If a nanostructure is introduced, local structures tend to be modified. The coordination number of atoms near the surface as well as shapes of coordination polyhedra must be changed. If the medium- and long-range interactions are cut, the forces acting on the surface atoms are also different from the bulk. These changes are caused by nano-sizing.

Some of these effects will be discussed in the following subsections. Other factors considered in nanoionics will be discussed in Chapters 3, 6, and 7.

### 2.6.2 Effects Related to the Surface Areas or Surface Regions and Possible Measures of Characteristics of Nano-Sizing

Let's consider the situation of one particle with different sizes. The ratio of the surface area and volume may be used as a measure to characterize nanoparticles. If we consider the case of cubic particles, the ratio is always $6\,l^2/l^3 = 6/l$, where $l$ is the side length. That is, the ratio is simply inversely proportional to the size of each cubic particle. An important problem to be considered here seems to be the thickness of the surface region. If it does not change with the particle size, the ratio of the surface region becomes larger and larger with nano-sizing. That is, in nano-sized particles, almost parts of the system have the characteristics as surfaces. If the thickness is 1 nm, it can be negligibly small compared with a large particle of >100 nm size, while in the particle of 10 nm size, the width cannot be negligible. Therefore, one needs to examine how large this thickness is and how it changes with size. Thus, one of the possible measures can be the length scale of the surface structure.

On the other hand, with nano-sizing, total surface areas and regions change drastically when density and total volume of the system are kept unchanged. For example, if one considers the cubic systems with side length $l_1$ = 1 nm and $l_2$ = 100 nm, with the same total volume $l_2^3$ = 1,000,000 (nm)$^3$, the total surface areas are 6,000,000 and 600 nm$^2$, respectively. That is, it increases with the number of partitions (with the power law with the exponent of 3). Therefore, if the phenomena depended on the surface areas only, such power law dependence or its function of the particle size might be expected. This large dependence is coming from the difference in dimensionality of the area and volume, that is the difference in the complexity of dimensions of space. If one considered a more complicated situation with fractal nature, a measure of the complexity is a fractal dimension. Thus, another possible measure of the nanostructured materials can be "fractal dimensions." Such dependence should also be modified if one considered the width of the surface regions and/or active regions characteristic to the phenomena.

If the properties of materials are related to the surface area, they will be affected considerably by the size of particles. One can easily imagine how large the difference in the solubility of powder sugar and sugar candy is. Melting is also affected by size. One can imagine the situation of snow and a large block of ice if one neglects the difference in phases.

In the field of pharmaceuticals, the dissolution process is known to be a function of the surface area [40] and such a process is important to control the dissolution of pharmaceuticals [41].

Adsorption is closely related to surface areas or regions and might be used to estimate the specific surface area. Experimentally, adsorption in porous media tends to be treated by Brunauer, Emmet, and Teller's (BET) methods [42] and used to measure the surface area. However, caution is required for the applicability of methods such as the BET equation in the case of nanostructured materials [43]. The limitation and the applicability of the method are also argued by using the binary Lennard–Jones model [44].

#### 2.6.2.1 Dissolution and nucleation of nanoparticles: Nernst–Noyes–Whitney equation and Noyes–Whitney equation

If the dissolution of a solid is governed by the diffusion obeying Fick's law, the process can be described by the following Nernst–Noyes–Whitney equation [40]:

$$dC/dt = DS/V\delta\ (C_s - C), \tag{2.3}$$

where $S$ is the surface area of the particle, $D$ is the diffusion coefficient, $V$ is the volume of solvent, $\delta$ is the thickness of the diffusive layer. $C_s$ is the saturated solubility of the particle in solution in the diffusion layer and $C$ the concentration of the particle. Here $D$ can be correlated with viscosity by the Stokes–Einstein equation, $D = RT/6\pi\eta rN$ for the case of sphere particles, with a radius $r$. If $D/(V\delta)$ is regarded as a constant $k$, the Noyes–Whitney equation is obtained as follows:

$$dC/dt = kS(C_s - C) \tag{2.4}$$

As found in this relation, when the particle size is reduced, the total effective surface area of a particle increases, and it results in the increase of the dissolution rate.

Of course, this equation is important for controlling the solubility of drugs. Furthermore, the nucleation theory is closely related to this process [45]. Namely, particle growth can be calculated using the Noyes–Whitney equation with the negative concentration gradient.

Again, caution is required for the applicability of such relations at the nanoscale. As pointed out in ref. [45], it may not work for the nanoscale. A similar situation is found in the self-healing process discussed in Chapter 7. When we consider the properties of nanoscale, they are often related to the events of both short and long time scales. For example, it was found that the self-healing process in porous silicates found in a short time scale region is not controlled by the motion in diffusive regimes. As shown in these examples, the applicability of such an equation in the nanoscale should be examined carefully.

### 2.6.3 Determination of Size and Strain of Nanoparticles by X-ray Diffraction Patterns

In nanostructured materials, nanoparticles and/or nano-sized domains are contained in many cases. They can be dispersed, mixed in layers, pellets, and films. In this situation, the determination of the particle size would be an important task in experimental approaches. There are several methods such as direct observations by SEM, determination of the (secondary) particle size, and distribution by dynamic light scattering (DLS). Since each method has its characteristics and errors, careful arguments are required.

In this subsection, the experimental method to determine the size of nanoparticles by X-ray diffraction will be explained briefly. The line breadths in the powder X-ray diffraction pattern are affected by both small particle size and strain broadening. In 1918, Scherrer proposed the method to determine the size of crystals from the width of the diffraction peak [46]. If other effects such as a strain of the system are negligibly small, mean crystallite sizes, $L$, can be determined from the broadening of the peak, $\beta$, by the following relation:

$$\beta_L = \frac{Kl}{L\cos\theta}, \tag{2.5}$$

where $K$ is a constant and $l$ is a wavelength, $\theta$ is a Bragg angle.

Here the constant $K$ depends on the definition of the size of the particle and that of the broadening. A careful choice of value is recommended [47]. On the other hand, if the contribution of the strain is considered, the following relation by Williamson and Hall is a good representation in many cases [48]:

$$\beta_{\text{total}}\cos\theta = C\varepsilon\sin\theta + \frac{Kl}{L}. \tag{2.6}$$

This equation was derived from Eq. (2.5) for the size broadening and the following Eq. (2.7) concerned with the strain broadening:

$$\beta_e = C\varepsilon\tan\theta \tag{2.7}$$

From the sum of (2.6) and (2.7),

$$\beta_{\text{total}} = \beta_e + \beta_L = C\varepsilon \tan\theta + \frac{Kl}{L\cos\theta}. \tag{2.8}$$

If both sides of the Eq. (2.8) are multiplied by cos $\theta$, observed for Eq. (2.6) can be obtained.

In Fig. 2.14, the Williamson–Hall plot observed for $BaF_2$ nanoparticles [49] is shown. In this case, results are found to depend on hydrothermal treatment times.

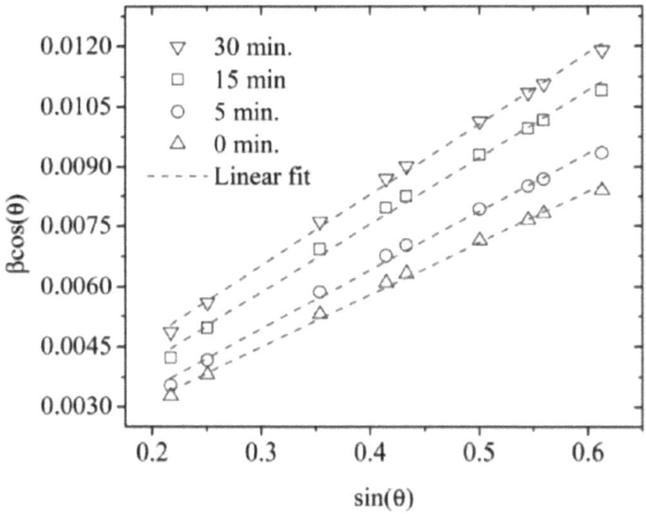


**Figure 2.14** Representative Williamson–Hall plots of the $BaF_2$ nanoparticles prepared using different hydrothermal treatment times. Reprinted with permission from Andrade, A. B., Ferreira, N. S., and Valerio, M. E. G. (2017), Particle size effects on structural and optical properties of $BaF_2$ nanoparticles, *RCS Adv.*, 43, pp. 26839–26848. The Royal Society of Chemistry, 2017.

As shown in this example, if $\beta_{\text{total}}\cos\theta$ is plotted against sin $\theta$, lines are obtained. From the slope, $C\varepsilon$ can be determined, while from the intersection, $Kl/L$ can be obtained. As shown in these relations, both strain and size are considered as the function of the lattice parameters. It means that nano-sizing of particles affects the lattice constant as well as the mechanical properties of materials.

New methods are still developing [50]. Comparisons of methods are found in several references [45, 51].

### 2.6.4 Effect of Changes of the Atomic Structure: Coordination Number and Formation of Different Phases

When the particle size is reduced, structures of materials are modified near the surface. For example, coordination number of the particle on the surface must be different from the bulk one. This situation can be found in some catalysts. Because the activity of the catalyst is affected by the coordination number [52], such changes should be considered.

The reduction of the size also causes the formation of different phases [53, 54]. For example, high activity of catalyst face-centered cubic (fcc) Ru nanoparticles has been reported by Kumara and coworkers [54]. Recently, it was reported that the Ru nanoparticle with the fcc structure confined in a porous cage (as-synthesized Ru NPs@PCC-2 composite) exhibited record-high catalytic activity in methanolysis of ammonia borane, which is important in chemical hydrogen storage [55].

## 2.7 Conclusion

Classification of the nanostructured materials is a useful method to treat them systematically and to design new nanostructured materials. A modified version of the classification of ref. [7] was proposed here. Then the classifications are expanded to more complex systems with fractal and/or multifractal characters. Porous systems are indexed, separately from the packed systems, although they can be treated as a function of density or porosity, which is the counter part of the packing factor. It is also possible to consider the fractal dimension of holes.

By using these substructures, all possible combinations can be considered. It will be beneficial to look for new combinations to form nanostructured materials.

Several effects of nano-sizing are discussed here and some methods to examine them are shown. However, one should note that the effects found in nanostructured materials are not necessarily explained by these effects. Further discussion of effects in nanostructured materials is continued in the following chapters.

## References


1. Habasaki, J., León, C., and Ngai, K. L. (2017), *Dynamics of Glassy, Crystalline and Liquid Ionic Conductors, Experiments, Theories, Simulations*, Springer International, Switzerland.
2. Habasaki, J. (2016), Molecular dynamics study of nano-porous materials—Enhancement of mobility of Li ions in lithium disilicate, *J. Chem. Phys.*, 145, 204503.
3. Habasaki, J., and Ngai, K. L. (2018), Heterogeneous dynamics in nanoporous materials examined by molecular dynamics simulations-effects of modification of caged ion dynamics, *J. Non-Crystal. Solids*, 498, pp. 364–371.
4. Habasaki J., and Ishikawa M. (2014), Molecular dynamics study of coagulation in silica-nanocolloid-water-NaCl systems based on the atomistic model, *Phys. Chem. Chem. Phys.*, 16(43), pp. 24000–24017.
5. Gleiter, H. (1995), Nanostructured materials: State of the art and perspectives, *Nanostruct. Mater.*, 6, pp. 3–14.
6. Gleiter, H. (2000), Nanostructured materials: Basic concepts and microstructure, *Acta Mater.*, 48, pp. 1–29.
7. Pokropivny, V. V., and Skorokhod, V. V. (2007), Classification of nanostructures by dimensionality and concept of surface forms engineering in nanomaterial science, *Mater. Sci. Eng., C*, 27, pp. 990–993. doi:10.1016/j.msec.2006.09.023; Pokpivny, V. V., and Skorohod, V. V. (2008), New dimensionality classification of nanostructures, *Physica E*, 40(7), pp. 2521–2525.
8. Liu, R., Duay J., and Lee, S. B. (2011), Heterogeneous nanostructured electrode materials for electrochemical energy storage, *Chem. Commun.*, 47, pp. 1384–1404.
9. Youn, H. C., Baral, S., and Fendler, J. H. (1988), Dihexadecyl phosphate, vesicle-stabilized and in situ generated mixed cadmium sulfide and zinc sulfide semiconductor particles: Preparation and utilization for photosensitized charge separation and hydrogen generation, *J. Phys. Chem.*, 92(22), pp. 6320–6327.

10. Dabbousi, B. O., Rodriguez-Viejo, J., Mikulec, F. V., Heine, J. R., Mattoussi, H., Ober, R., Jensen, K. F., and Bawendi, M. G. (1997), (CdSe)ZnS core–shell quantum dots: Synthesis and characterization of a size series of highly luminescent nanocrystallites, *J. Phys. Chem. B,* 101, pp. 9463–9475.
11. Chaudhuri, R. G., and Paria, S. (2012), Core/shell nanoparticles: Classes, properties, synthesis mechanisms, characterization, and applications, *Chem. Rev.*, 112, pp. 2373–2433.
12. Ben-Avraham, D., and Havlin, S. (2000), *Diffusion and Reactions in Fractals and Disordered Systems*, Cambridge University Press, Cambridge.
13. Mandelbrot, B. (1967), How long is the coast of Britain? statistical self-similarity and fractional dimension, *Science*, 156, pp. 636–637.
14. Chhabra, A., and Jensen, R. V. (1989), Direct determination of the $f(\alpha)$ singularity spectrum, *Phys. Rev. Lett.*, 62, pp. 1327–1330.
15. Halsey, C., Jensen, M. H., Kandanoff, L. P., Procaccia, I., and Shraiman, B. I. (1986), Fractal measures and their singularities: The characterization of strange sets, *Phys. Rev. A*, 33, pp. 1141–1150.
16. Berger, J. M., and Mandelbrot, B. B., Multifractals and $1/f$ Noise: Wild Self-Affinity in Physics (1963–1976) (Selecta; V.N).
17. Sorensen, C. M., Kim, W., Fry, D., Shi, D., and Chakrabarti, A. (2003), Observation of soot superaggregates with a fractal dimension of 2.6 in laminar acetylene/air diffusion flames, *Langmuir*, 19, pp. 7560–7563.
18. Fry, D. Chakrabatti, A., Kim, W., and Sorensen, C. M. (2004), Structural crossover in dense irreversibly aggregating particulate systems, *Phys. Rev. E*, 69, 061401(1–10).
19. Tiwari, J. N., Tiwari, R. N., Kim, K. S. (2012), Zero-dimensional, one-dimensional, two-dimensional and three-dimensional nanostructured materials for advanced electrochemical energy devices, *Prog. Mater. Sci.*, 57, pp. 724–803.
20. Deer, W. A., Howie, R. A., Wise W. S., and Zussman, J. (2004), *Rock-Forming Minerals. Volume 4B. Framework Silicates: Silica Minerals. Feldspathoids and the Zeolites,* 2nd ed., London, Bath, Geological Society of London.
21. Henderson, G. S. (2005), The structure of silicate melts: A glass perspective, *Can. Mineralogist*, 43, pp. 1921–1958.
22. Habasaki, J., and Ngai, K. L. (2008), Heterogeneous dynamics of ionic liquids from molecular dynamics simulations, J. Chem. Phys., 129, 194501 (1–16).

23. Greaves, G. N., and Sen, S. (2007), Inorganic glasses, glass-forming liquids and amorphizing solids, *Adv. Phys.*, 56, pp. 1–166.
24. Doweidar, H. (1996), The density of alkali silicate glasses in relation to the microstructure, *J. Non-Cryst. Solids*, 194, pp. 155–162. https://doi.org/10.1016/0022-3093(95)00489-0.
25. Habasaki, J., and Ngai, K. L. (2013), Molecular dynamics study of network statistics in lithium disilicate: $Q_n$ distribution and the pressure-volume diagram, *J. Chem. Phys.*, 139, 064503 (1–14).
26. Wong-Foy, A. G., Lebel, O., and Matzger, A. J. (2007), Porous crystal derived from a tricarboxylate linker with two distinct binding motifs, *J. Am. Chem. Soc.*, 129, pp. 15740–15741.
27. Koh, K., Wong-Foy, A. G., and Matzger, A. J. (2009), A porous coordination copolymer with over 5000 m2/g BET surface area, *J. Am. Chem. Soc.*, 131, pp. 4184–4185.
28. Yang, X.-Y., Chen, L.-H., Li, Y., Rooke, J. C., Sanchez, C., and Su, B.-L. (2017), Hierarchically porous materials: Synthesis strategies and structure design, *Chem. Soc. Rev.*, 46, pp. 481–558.
29. Stauffer, D. (1985), *Introduction to Percolation Theory*, Taylor & Francis, London and Philadelphia, and references therein.
30. Allain, C., and Cloitre, M. (1991), Characterizing the lacunarity of random and deterministic fractal sets, *Phys. Rev. A*, 44, pp. 3552–3558.
31. Sing, K. S. W., Everett, D. H., Haul, R. A. W., Moscou, L., Pierotti, R. A., Rouquerol, J., and Siemieniewska, T. (1985), Reporting physisorption data for gas solid systems with special reference to the determination of surface-area and porosity (Recommendations 1984), *Pure Appl. Chem.*, 57, pp. 603–619.
32. Rouquerol, J., Avnir, D., Fairbridge, C., et al. (2019), Recommendations for the characterization of porous solids (Technical Report), *Pure Appl. Chem.*, 66, pp. 1739–1758. doi:10.1351/pac199466081739.
33. Adler, P. M., and Thovert, J. F. (1993), Fractal porous media, *Transp. Porous Med.*, 13, pp. 41–78. https://doi.org/10.1007/BF00613270.
34. O'Neill, M. L., Haas, M. K., Peterson, B. K., Vrtis, R. N., Weigel, S. J., Wu, D., Bitner, M. D., and Karwacki, E. J. (2006), Impact of pore size and morphology of porous organosilicate glasses on integrated circuit manufacturing, *Mater. Res. Soc. Symp. Proc.* vol. 914-F01-02, © 2006 Materials Research Society.
35. Hashin, Z., and Shtrikman, S. (1962), A variational approach to the theory of the elastic behaviour of polycrystals, *J. Mech. Phys. Solids*, 10, pp. 343–352.

36. Hashin, Z., and Shtrikman (1963), A variational approach to the theory of the elastic behaviour of multiphase materials, *J. Mech. Phys. Solids*, 11, pp. 127–140.

37. Yu, B., and Li, J. (2001), Some fractal characters of porous media, *Fractals*, 9, pp. 365–372.

38. Maslov, V. P. (2007), General notion of a topological space of negative dimension and quantization of its density. *Mathematical Notes*. 81, 140. doi:10.1134/S0001434607010166.

39. Leslie-Pelecky, D. L., and Rieke, R. D. (1996), Magnetic properties of nanostructured materials, *Chem. Mater.*, 8, pp. 1770–1783.

40. Noyes, A. A., and Whitney, W. R. (1897), Ueber die Auflösungsgeschwindigkeit von festen Stoffen in ihren eigenen Lösungen, *Z. Phys. Chem.*, 23, pp. 689–692. Published Online: 2017-01-22 | DOI: https://doi.org/10.1515/zpch-1897-2341.

41. Dizaj, S. M., Vazifehasl, Z. H., Salatin, S., Adibkia, K. H., and Javadzadeh, Y. (2015), Nanosizing of drugs: Effect on dissolution rate, *Res. Pharm. Sci.*, 10(2), pp. 95–108.

42. Brunauer, S., Emmett, P. H., and Teller, E. (1938), Adsorption of gases in multimolecular layers, *J. Am. Chem. Soc.*, 60, pp. 309–319. doi:10.1021/ja01269a023. ISSN 0002-7863.

43. Rouquerol, J., and Llewellyn, P., and Rouquerol, F. (2007), Is the BET equation applicable to microporous adsorbents? *Studies in Surface Science and Catalysis*, 160, pp. 49–56. 10.1016/S0167-2991(07)80008-5.

44. Gelb, L. D., and Gubbins, K. E. (1998), Characterization of porous glasses: Simulation models, adsorption isotherms, and the Brunauer-Emmett-Teller analysis method, *Langmuir*, 14, pp. 2097–2111.

45. Sugano, K. (2012), *Biopharmaceutics Modeling and Simulations: Theory, Practice, Methods, and Applications*, John Wiley & Sons.

46. Scherrer, P. (1918), Bestimmung der Größe und der inneren Struktur von Kolloidteilchen mittels Röntgenstrahlen, *Nachr. Ges. Wiss. Göttingen*, 26, pp. 98–100.

47. Saito, Y. (1975), *X-Ray Crystallography* (last volume in set) 5th edition, Chapter IV-2, Maruzen, in Japanese.

48. Williamson, G. K., and Hall, W. H. (1953), X-ray line broadening from filed aluminium and wolfram, *Acta Metallurgica*, 1(1), pp. 22–31.

49. Andrade, A. B., Ferreira, N. S., and Valerio, M. E. G. (2017), Particle size effects on structural and optical properties of $BaF_2$ nanoparticles, *RCS Adv.*, 43, pp. 26839–26848.

50. He, B. B. (2003), Introduction to two-dimensional X-ray diffraction, *Powder Diffraction*, 18, pp. 71–85.

51. Dorofeeva, G. A., Streletskiib, A. N., Povstugara, I. V., Protasova, A. V., and Elsukova, E. P. (2012), Determination of nanoparticle sizes by X-ray diffraction, *Colloid J.*, 74, pp. 675–685.

52. Liu, P., and Zheng, N. (2018), Coordination chemistry of atomically dispersed catalysts, *Natl. Sci. Rev.*, 5, pp. 636–638.

53. Michalska, Z. M., and Webster, D. E. (1974), Supported homogeneous catalysts, *Platinum Met. Rev.*, 18, pp. 65–73.

54. Kumara, L. S. R., Sakata, O., Kohara, S., Yang, A., Song, C., Kusada, K., Kobayashi, H., and Kitagawa, H. (2016), Origin of the catalytic activity of face-centered-cubic ruthenium nanoparticles determined from an atomic-scale structure, *Phys. Chem. Chem. Phys.*, 18, pp. 30622–30629.

55. Fang, Y., Li, J.-L., Togo, T., Jin, F.-Y., Xiao, Z.-F, Liu, L.-J, Drake, H., Lian, X.-Z., and Zou, H.-C. (2018), Ultra-small face-centered-cubic Ru nanoparticles confined within a porous coordination cage for dehydrogenation, *Chem*, 4, pp. 555–563.

# Chapter 3

# Nanostructures in Nanoionics and Colloidal Chemistry: Overview and Problems

This chapter consists of two parts. In the first half of this chapter, some experimental facts for the problems related to enhanced ion dynamics in porous and related materials are summarized along with the history of research of the LiI–alumina composite. Historical view of how the research in nanoionics has been developed will be useful to understand what factors affect the problems and what kind of methods can be used. Some of the results seem to be contradicting each other so that further research and methods are required for the clarification of the mechanism.

In the second half of this chapter, experimentally relevant facts for the structures obtained from the coagulation of colloids/particles will be summarized, where the importance of dimensionality of the systems in colloidal systems and gels is shown. Furthermore, evidence of the coexistence of several length scales in gels, which means the existence of super aggregates, will be presented.

*Molecular Dynamics of Nanostructures and Nanoionics: Simulations in Complex Systems*
Junko Habasaki

ISBN 978-981-4800-77-8 (Hardcover), 978-1-003-04490-1 (eBook)
www.jennystanford.com

## 3.1 Historical and Current Research on Nanoionics

The first topic is to show the historical and current status in the field of nanoionics. Some experimental facts and related simulation works so far reported will be explained. Especially, attention is paid to nanoionics with enhanced dynamics and factors modifying ionic properties in nanostructured composites for suggesting problems to be examined by molecular dynamics simulations.

Characteristics of nanostructured materials tend to be explained by assuming them to be coming from their large surface area as well as the large boundary regions. However, there are many other possible causes of changes in properties in nanostructured materials [1]. Furthermore, possible roles of large surface area and/or grain boundaries are not simple.

After the first report by Liang [2] for enhanced dynamics in $LiI–Al_2O_3$ composites, similar findings in other composites were reported by many authors. These findings have brought developments of studies of "surfaces and boundaries," "chemical preparations of composites," as well as developments of "percolation theories." Therefore, the results of these studies will be useful to learn the history and current status of the field of nanoionics and related ones, although the explanation is not exactly in the chronological order. Many kinds of explanations have been proposed so far, regardless that quite similar behaviors of systems are commonly observed. One of the origins of this complexity seems to be coming from the dual (or more) roles of each component. For example, alumina and related components treated here are sometimes regarded as low-conductivity materials or insulators to separate conductive phases. However, at the same time, they are porous supports to hold and affect the conducting phases. Another possible cause may be the complexity of the experimental conditions to prepare and examine nanostructured materials. Each method seems to have its own characteristics to be considered to conclude.

For understanding dynamics in composites, methods to distinguish such roles seem to be required. Systematic studies for the role of modification of the system will be also helpful. Several proposals to explain the enhancement and distinguish several origins will be explained in this chapter. The effects of modification of dynamics in porous systems are discussed in Chapters 6 and 7. In Chapter 6, methods to separate possible mechanisms and effects of introducing pores by using MD simulations will be shown for the case of porous silicate systems. The methods used can be applied to more complicated nanostructured systems such as composites and films, confined systems with/without some modifications.

### 3.1.1 Experimental Observation for LiI–Alumina Composite

As already mentioned, the first important experimental facts to be argued in this field is the existence of enhanced dynamics found in several composites or mixtures. In 1973, Liang [2] reported enhanced dynamics of ions in the lithium iodide–aluminium oxide (LiI-alumina) system.

Liang [2] reported that the composite of lithium iodide–aluminium oxide showed the maximum conductivity of ~$10^{-5}$ (S $cm^{-1}$) at around 45 mol % of $Al_2O_3$ content at room temperature, as shown in Fig. 3.1. After the discovery, this and related systems have been examined by many researchers [3–19].

Many theories and models have been proposed to explain enhancement. The space charge layer model [5–7], deformation-induced defect model [8–11], the effect of water [13, 14] and percolation theories [15, 16] are important ones, which affects developments of this research field. Other possible explanations can be found in our book of ionics, especially in the chapters of nanoionics and further applications [1]. Temperature dependence reported in [3] suggested that the enhancement is remarkable when it is concerned with the cooperative motions of ions at higher temperatures.

These explanations contain several factors and are not necessarily independent of each other.

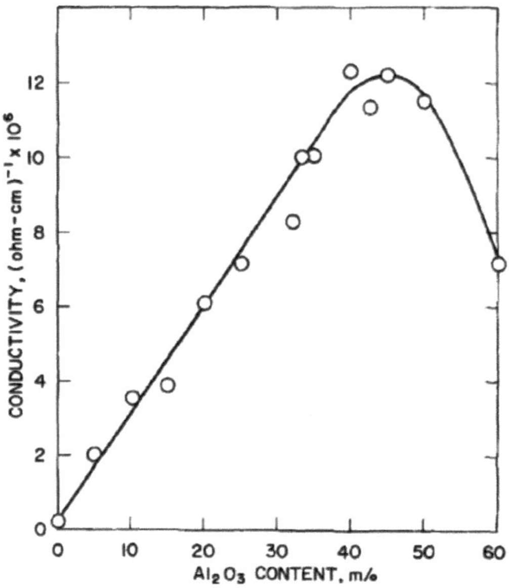


**Figure 3.1** Conductivity of LiI–$Al_2O_3$ composite as a function of alumina content. Reprinted with permission from Liang, C. C. (1973), *J. Electrochem. Soc.*, 120(10), pp. 1289–1292, doi:10.1149/1.2403248. Copyright (1973) Elsevier.

### 3.1.2 Space-Charge Layer Model-Geometrical Factors Related to Ionic Paths

In the space-charge layer model by Wagner [5], and that by Maier [6, 7], cations are trapped at the interfaces of two phases and cation vacancies increase within the space-charge layer due to a charge balance.

The concept of the model seems to be related to the percolation theory because the conduction path formed by the space-charge layer is affected by insulators introduced. The model seems to work well for several systems. Changes in the length scale may be expected from the model, and the method to obtain Debye length discussed in Section 3.7 is applicable for this purpose. Therefore, it will be interesting to examine the structure of the space-charge layers including length scales of it by MD simulations.

### 3.1.3 Deformation-Induced Defects Model

Although the space-charge model can explain the experimental fact in several cases, Dudney [9–11] reported that the space-charge layer could not explain successfully the enhancement of dynamics and proposed that the deformation-induced defects stabilized by inert phases cause the enhanced conductivity. It was found that the composites of AgCl and $Al_2O_3$ prepared by impregnation of an alumina fiber paper with molten AgCl did not show any enhancement, while the deformed one showed the enhancement [10]. It was also pointed out that the samples exhibiting enhanced conductivity so far reported were prepared as a pellet under pressure.

It is worth mentioning that one must consider the effects of pressure and tension remaining in the system.

The possible mechanism caused by deformation seems to be concerned with not only geometrical factors of paths but also the structure of cages discussed in Chapters 6 and 7.

### 3.1.4 Effect of Sizes of Particles in LiI–Alumina Composite

In 1979, Stoneham and coworkers [4] demonstrated that lithium iodide shows enhanced ionic conductivity when doped with a powder of the insulator, alumina. They extended Landauer's effective medium model [17] for electric resistance of binary systems to see if the observations were consistent with a high-conductivity layer forming on each non-conducting particle. They introduced a model, where the conducting region does not end abruptly but tails off into the bulk, and the model worked well. They have argued how the size of particles of powder affects the conductivity when high-conductivity layer surrounding insulator particles exists and has shown that the existence of the maximum of the enhancement. They suggest that the highest ionic conductivity will be reached with very fine alumina powder so that blockage is minimized. The effects of the size of particles seem to modify the geometrical factors related to ionic paths.

### 3.1.5 Existence of Distinct Sites

Ardel and coworkers [3] have examined the mixture of LiI and particles of $Al_2O_3$, experimentally. In this case, the system is prepared as the pellet treated under the press. Characterization of the system was done by several experimental methods such as X-ray diffractions, complex impedance, and high-resolution $^7Li$ NMR spectroscopy. From these experiments, it was found that the observed enhancement in ionic conductivity by the addition of $Al_2O_3$ is accompanied with an increase in grain boundary resistance. By the high-resolution (MAS) NMR method, the existence of two or more distinct $Li^+$ sites was found. It was also found that populations of different kinds of sites depend on the composition of mixing and temperature. One was assigned to the characteristic of bulk LiI and the rest to the surface/interface regions. Ionic conductivity is thus explained related to their populations. They also reported that the increase of $Al_2O_3$ concentration from 0 % to 50 % at temperatures below 90 °C is followed by a conductivity increase of only about 60 %, while above 90 °C conductivity enhancement of over a factor of two. They have found that at low temperatures the effect of alumina content on ionic conductivity in LiI–$Al_2O_3$ is less pronounced.

### 3.1.6 Percolation Model

Percolation theory can treat situations where conduction paths are separated by boundaries or pores.

Several versions of percolation models have been proposed so far [15, 16].

In Fig. 3.2, a three-component bond-percolation model by Roman and coworkers [15] is shown. A remarkable feature of this model is the existence of two threshold concentrations, and the model can explain the existence of the maximum of the conductivity at a certain concentration.

In the figure, the two-phase mixture is represented on a square lattice for different concentrations, $p$, of the insulating material, which is represented by the shaded regions. In this figure, the highly conducting bonds are marked by bold lines. In Fig. 3.2a, the concentration, $p$, of the insulator is smaller than $p_c'$ so that the path on the interface is not percolated. In the case of Fig. 3.2b, $p$ is comparable to $p'_c$, which is the onset of interface percolation. In Fig. 3.2c, $p$ is comparable to the second threshold $p''_c$ for disruption of the conducting paths. In Fig. 3.2d, the value $p$ is larger than $p''_c$, and therefore portions of insulators spread widely.

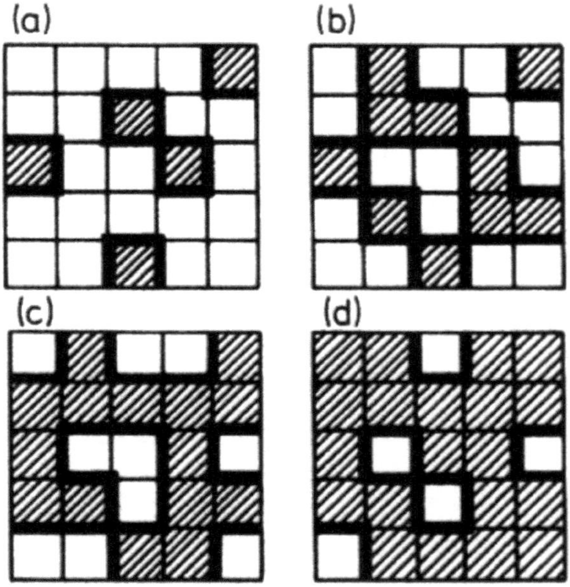


**Figure 3.2** The two-phase mixture on a square lattice for different concentration $\boldsymbol{p}$ of the insulating material, represented by the shaded regions. The highly conducting bonds are marked by bold lines. (a) $\boldsymbol{p} < \boldsymbol{p}'_c$ (b) $\boldsymbol{p} = \boldsymbol{p}'_c$ onset of interface percolation, (c) $\boldsymbol{p} = \boldsymbol{p}''_c$ threshold for disruption of the conducting paths, and (d) $\boldsymbol{p} > \boldsymbol{p}''_c$. Reprinted with permission from Roman, H. E., Bunde, A., Dieterich, W. (1986), Conductivity of dispersed ionic conductors: A percolation model with two critical points, *Phys. Rev. B*, 34, pp. 3439–3445.

Although the model, which includes several components, can reproduce the behavior similar to experiments, direct comparison with microscopic behavior of the system seems to be desired and molecular dynamics simulations seem to have the power to bring more insights for such behaviors.

Heitjans and Wilkening [18] have shown $^7$Li nuclear magnetic resonance spectra of nanocrystalline $(1-x)Li_2O$: $xAl_2O_3$ composites with different amounts ($x$) of alumina as shown in

Fig. 3.3. They have pointed out that a larger area fraction of the mobile ions was found when a higher number of conductor–insulator contacts concerning the total $Li_2O$ content were found. The result can be explained also by the percolation model. The existence of the localized fast ionic motion is considered in that figure.

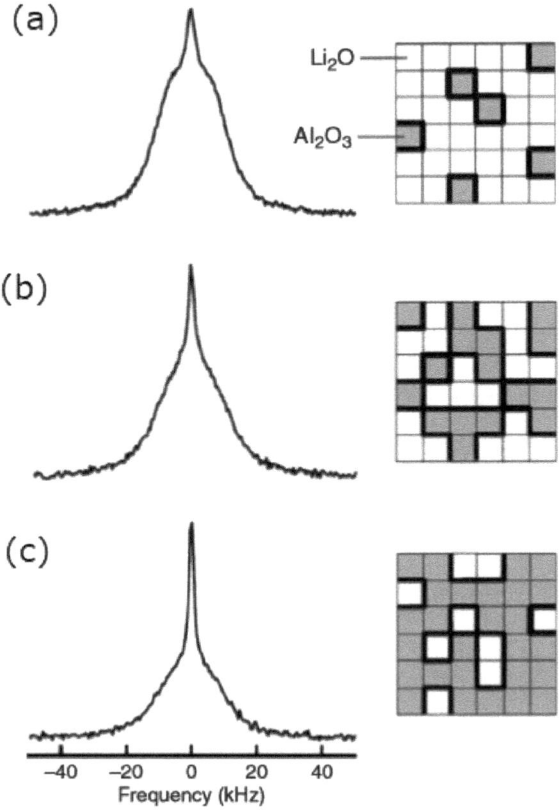


**Figure 3.3** $^7$Li nuclear magnetic resonance spectra of nanocrystalline (1–$x$) $Li_2O$: $x$ $Al_2O_3$ composites with different amounts ($x$) of alumina (a) $x$ = 0.2, island-like heterostructure, where small volume fraction is for alumina. (b) $x$ = 0.5, where percolating diffusion pathways are formed and it corresponds to the dc-conductivity maximum. (c) $x$ = 0.8, where isolated islands of $Li_2O$, which is fast but localized pathways are formed. Spectra were recorded at 433 K and 58.8 MHz. Reprinted with permission from Heitjans, P., Wilkening, M. (2009), Ion dynamics at interfaces: Nuclear magnetic resonance studies, *MRS Bull.*, 34, pp. 915–922, with some modifications.

### 3.1.7 A Length Scale Related to Enhanced Dynamics

Schreck and coworkers [19] have studied the enhanced ionic conductivity of thin films of LiI evaporated onto a planar sapphire (sapphire is an alumina oxide ($\alpha$-$Al_2O_3$), which usually containing trace amounts of elements such as Ca, Fe) surface. It was found that the specific conductivity of LiI adjacent to the sapphire film (with a thickness of 350 Å) exceeded the bulk value by nearly one order. With distance from the surface, the value decreases exponentially. The conductivity of the film varies with temperature (25 °C $\leqq T \leqq$ 120 °C) following activation energy of 0.40±0.04 eV. They argued that the value is in good agreement with the activation energy of bulk LiI in the extrinsic region. The corresponding "Debye-length," over which the conductivity drops to $1/e$ of its maximum value at the interface, was found to be about 300 Å.

The Debye-length seems to be longer than the typical MD basic cell size, and hence one should be careful to design the MD simulations of such systems.

### 3.1.8 Effect of Impurity Such as Water

It has been reported that the enhancement is affected by how to prepare the sample as shown in the following examples. The effects of impurity, especially water brought during the experimental procedures, have been examined by several works.

Skarstad and coworkers [12] have compared $LiI{\cdot}H_2O$ and LiI to learn if the enhancement is due to the presence of water. Measurements of the conductivity of $LiI{\cdot}H_2O$ give values $\sim 10^{-5}$ S cm$^{-1}$, while, higher conductivities, $10^{-4}$ S cm$^{-1}$, are observed for LiI and $LiI{\cdot}H_2O$ mixed with $SiO_2$ or $Al_2O_3$. It means that the conductivity of the monohydrate is not high enough to explain the enhancement. That is, the effect cannot be explained by the presence of water alone.

This problem was also examined by Bjorkstam and coworkers [13] by comparing single crystals of $LiI{\cdot}D_2O$ and $LiI{\cdot}H_2O$, as well as upon pure and $SiO_2$ doped $LiH{\cdot}H_2O$ powder. In that work, nuclear magnetic resonance studies of line narrowing ($T2$), spin-lattice relaxation ($T1$), and spectral changes with temperature have been performed to measure activation energies and correlation times for $Li^+$ ion motion and water molecule reorientation and these results are compared with those from conductivity and structural experiments. The effect of water is found to be non-negligible for the enhancement.

### 3.1.9 Effect of Water in Related Systems

Enhancement is widely observed in related systems. The effects of water mentioned in the previous section are also observed in $Li_xMn_2O_4$. Chen and coworkers [14] have examined $Li_xMn_2O_4$ with $0.5 \leq x \leq 2.5$. It was found that the lithium chemical diffusion coefficient (chemical diffusion coefficient is for diffusion under concentration gradient) in $Li_xMn_2O_4$ is fairly affected by the preparation method. If it was synthesized by this wet-chemical technique, at least 10 times higher value than that of $Li_xMn_2O_4$ was obtained by the solid-state reaction method. Comparing the enhancement by wet and dry methods, the effect of water on the enhancement is suggested.

The maximum of the conductivity and the smaller slopes in the Arrhenius plots are commonly found in some composites of AgI and $M_xO_y$ (=$Al_2O_3$, $ZrO_2$, and $SiO_2$) with different compositions [20]. Comparable behaviors are also observed for nanocrystalline $BaF_2$ and $CaF_2$ and in $BaF_2$:$CaF_2$ composites [21].

There are many trials to understand the origin of the enhancement. Pre-factor and activation energy found in the Arrhenius plot seems to be useful to distinguish the possible mechanisms. Phipps and Whitmore [22] have observed the smaller (about 1/3 of bulk LiI) activation energy of the LiI sandwiched by $SiO_2$ plates as well as the increase of the pre-factor.

Yamada and coworkers [23] have examined the LiI–$SiO_2$ system by XRD, geometric density, and a.c. impedance. They concluded that their result with nearly constant activation energy suggested that ionic conductivity was enhanced by not deformation-induced defects but space-charge layer at interfaces. They found that the maximum ionic conductivity of $5.8 \times 10^{-6}$ S $cm^{-1}$ was observed when the amount of LiI was slightly smaller than the amount to fill pores. In that case, the maximum conductivity was obtained for the composite with the volume fraction of $SiO_2$ of 0.173. Although these works seem to support the existence of enhancement at the surface of some materials, when the adjacent layer of

LiI exists, the mechanism in different situations, further research, including the trapping effect in mesoporous systems and different sizes of pores, seems to be necessary to conclude the mechanism.

### 3.1.10 Lithium Silicate Glass Grown within Pores and in Composites

Similar enhancement of dynamics was observed in silicate glass within porous materials. Especially, systems containing lithium ions are important for designing high-conductivity materials towards solid-state batteries. In Section 3.1.6, alumina, silica, and other network formers are just considered as insulators to separate the ionically conducting phases, and hence the changes in the structures for the ionic conducting phase are not considered clearly. Different from these approaches, recently, the role of the free space in porous materials has been focused on in several papers.

### 3.1.11 Lithium Silicate in Porous Systems

For example, Saha and coworkers [24] have examined the conductivity of the system $23Li_2O{\cdot}77SiO_2$ in a pellet formed by ZnO nanorods. In this case, silicate glass was grown within the pores of a pellet formed by ZnO nanorods of a diameter of 70 nm, where lengths are changed from 80 to 270 nm. Remarkable enhancement of conductivity is found in such conditions. That is, nanoglass supported in nanorods shows three orders of lesser resistivity and 35 % lower activation energy compared to its bulk counterpart. This effect was ascribed to an increase in free volume in the nano-dimensional glass. The ac conductivity variation at high frequency showed an exponent of 0.5 indicating two-dimensional movements of lithium ions in the nanoglass.

This effect of introducing pores by support materials is closely related to the results of molecular dynamics simulations of porous lithium silicate, which will be discussed in Chapters 6 and 7. The effect caused by free volume seems to be concerned with two factors: vacancies on the jump paths of ions and the caging of ions. The important difference between these factors is as follows. The former factor is mainly concerned with the geometrical conditions related to the jump paths, while in the latter, the deformation or introducing pores affects the jump rates of the of ions through structures of cages. To distinguish the mechanism by the geometry of jump paths and changes in caged dynamics, it is convenient to consider the time-dependent behavior of the mean squared displacement of ions (or corresponding frequency dependence of the conductivity spectra). That is, these factors can be distinguished by the time scale of events as shown in Chapters 6 and 7. In both cases, the roles of cooperative motions should be taken into account.

### 3.1.12 Ion Transport in the Dilithium Ethylene Dicarbonate ($Li_2EDC$)

A complex system including Li ions has been studied by Borodin and coworkers [25]. That is, ion transport in the dilithium ethylene dicarbonate ($Li_2EDC$) component of the solid electrolyte interphase was examined by impedance spectroscopy and molecular dynamics (MD) simulations. When the melt is compared with the ordered system, the absolute value of the slope of the latter in Fig. 3.4 is smaller. That is, a decrease in the activation energy is observed. Snapshots of the structures of melt and the ordered system are shown in Fig. 3.4 and a clear ordering of Li ions is observed in Fig. 3.4c.

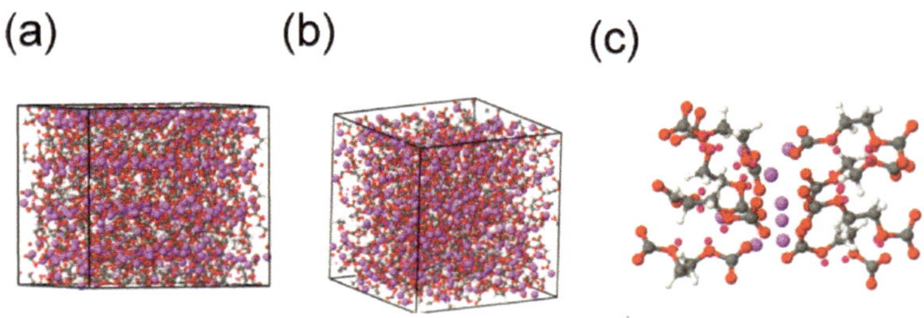


**Figure 3.4** Snapshot from the MD simulations of $Li_2EDC$ of the (a) ordered and (b) amorphous (molten phase) states at 450 K ($Li^+$ are highlighted as pink balls). A representative cluster from bulk $Li_2EDC$ (ordered) with distances less than 2.4 Å connected by dashed lines is also shown (c). Reprinted with permission from Borodin, O., Zhuang, G. V., Ross, P. N., Xu, K. (2013), Molecular dynamics simulations and experimental study of lithium ion transport in dilithium ethylene dicarbonate, *J. Phys. Chem. C*, 117(15), pp. 7433–7444. Copyright (2013) American Chemical Society.

From such aligned structures of Li ions, it may be expected that the enhancement of ion dynamics is caused by the cooperative motions of ions [26].

Conductivity from MD simulations using a polarizable model for melts and ordered $Li_2$EDC materials is shown in Fig. 3.5. As shown in this figure, when temperature decreases, the $Li^+$ diffusion coefficient between $Li^+$ layers in the ordered sample decreases slower than the diffusion coefficient of melts.

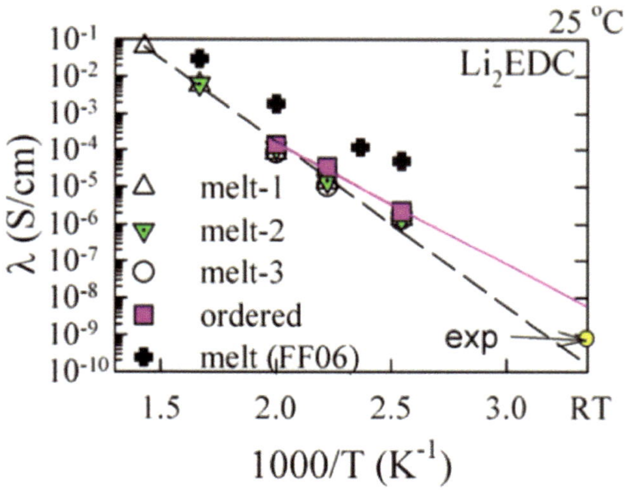


**Figure 3.5** Conductivity from MD simulations using a polarizable APPLE&P force field for melts and ordered $Li_2$EDC materials. Results from the MD simulations using a previous force field FF06 are also shown for comparison (crosses). $\Delta E$ (melt) = 84 kJ/mol, $\Delta E$ (ordered) = 64 kJ/mol. Reprinted with permission from Borodin, O., Zhuang, G. V., Ross, P. N., Xu, K. (2013), Molecular dynamics simulations and experimental study of lithium ion transport in dilithium ethylene dicarbonate, *J. Phys. Chem. C*, 117(15), pp. 7433–7444. Copyright (2013) American Chemical Society.

At low temperatures, the $Li^+$ diffusion across the layers is expected to become negligibly small compared to the diffusion within the $Li^+$ layers. Heterogeneous and non-Gaussian motions of lithium ions have been observed at 393 K. The difference between polarizable and non-polarizable models has also been discussed. Heterogeneous dynamics in other ionic systems related to these results will be discussed in Chapter 5.

### 3.1.13 Ion Transport in Polymer Electrolyte–$SiO_2$ Nanospheres in Poly(Ethylene Oxide)

Polymer electrolyte with silica has been examined by Lin and coworkers [27]. They measured conductivities of some composite solid polymer electrolyte. Monodispersed $SiO_2$ nanospheres in poly(ethylene oxide) showed good ionic conductivity ($1.2 \times 10^{-3}$ S cm$^{-1}$ at 60 °C, $4.4 \times 10^{-5}$ S cm$^{-1}$ at 30 °C), as well as electrochemical stability of the window up to 5.5 V. They have concluded that these systems show excellent rate capability as well as good cycling performance applicable to all-solid-state lithium batteries.

For a better understanding of the mechanisms of enhancements, following argument will be useful. Explanations so far proposed for the mechanism of enhancement seem to be related to the following factors.

Structural changes of ionic paths tend to be caused by the mixing of low conducting parts or insulator. In many papers, alumina or silica in composites is regarded as an insulator, which affects the paths of ions. These paths are affected by geometrical factors such as the structures of boundaries, layers, particle sizes, vacancies, and disorders.

Another possibility is that structural changes of cages are caused by introducing pores. It is because the insulators such as alumina or silica used in composite showing enhanced dynamics have a network structure, which is regarded as the porous support of ions in the ionic compounds (see Chapters 6 and 7).

### 3.1.14 Separation of Possible Factors Related to the Enhancement

Besides the factors mentioned in the previous sections, there are several effects found in nanostructured materials. It should not be forgotten that the difference of order and disorder of the structure in the system has a large effect (which can be several orders in magnitude) on the diffusivity and/or conductivities of ions [28–30]. Effects can be caused by the specific chemical or physical structure of materials.

Which factor (or factors) is (are) the most relevant among many possible factors? How can we distinguish the effect from others?

In the following chapters, such problems will be treated by MD simulations.

Here we briefly mention the outline of Chapters 4–7 to understand the relations among the chapters.

The dynamics in ionically conducting materials are complicated ones, which often deviate from Gaussian dynamics in several kinds of observations. Therefore, first, to examine the motions of ions in typical ionic conductors will be useful. This will be shown in Chapter 5. The characteristics of ionic motions explained here are used as the standard system to understand the effects of further modifications of systems.

In Chapter 5, from the results of MD simulations of lithium metasilicate glasses, fundamentals of ionics, including time scales and length scales found in the mean squared displacement in lithium silicate systems, will be explained. It is found that ionic motion is strongly heterogeneous and fast ions are accelerated by both short intervals of jumps as well as strong forward correlated motions in ion channels. The acceleration by the shortened time intervals of jumps is caused by the loosening of cages in a short time and length scales. While the forward correlated motions of ions are affected by the cooperativity of motions. As easily imagined, the latter motion is concerned with the dimensionality of the paths in longer time and length scales.

As shown in Fig. 3.6, the motion is accelerated by cooperative jumps of several Li ions and such motion is characterized [26] as Lévy flight dynamics [31]. Such kind of acceleration by cooperative motions is commonly observed in recent MD simulation of fast ion conductors [32–35]. The trajectories of ions form the shape of ion channels as already shown in Fig. 2.9 in Chapter 2. That is, the changes of motions and paths (ion channels) related to the percolation theory are one of the possible explanations. This kind of motion can also play roles in other composites. For instance, the alignment of Li ions in Fig. 3.5c seems to be suitable for this kind of motions, although further confirmation is necessary.

How can we distinguish two mechanisms relevant to the acceleration of dynamics?

In Chapter 6, the effects of modification observed in a different time and length scale regions will be shown for porous lithium silicate systems. The treatment of these systems can be a good example of how many possible effects caused by perturbations of systems can be distinguished by using MD simulations. As shown in the experimental facts for lithium-containing systems shown above, enhancement of ionic motion is quite common in many systems. Therefore, it is probable that the comparable mechanism observed in porous lithium silicate found in MD simulations in the *NVE* condition can contribute to the enhanced dynamics in related composites. As shown in this example, we suggest that one of the possible methods to confirm the mechanism is to examine the time scales and length scales of events concerning the enhancement.

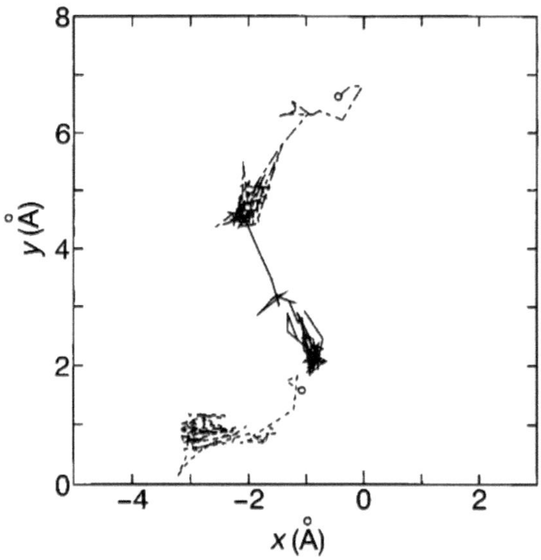


**Figure 3.6** Trajectories of the lithium ions showing cooperative jumps projected on the ***x-y*** plane during 16 ps at 700 K in lithium metasilicate. Reprinted from Habasaki, J., Okada, I., Hiwatari, Y. (1997), Fracton excitation and Lévy flight dynamics in silicate glasses, *Phys. Rev. B*, **55**, pp. 6309–6315 with permission. Copyright (1997). American Physical Society.

Since the dynamics of ions are known to be heterogeneous in glasses or crystals, it is interesting to consider the effects of modification of such characteristics in the porous systems as well as in composites. The characteristics of non-Gaussian (NG) dynamics shown in Section 3.2.2 are like those found in the single alkali systems [36, 37]. Modification of non-Gaussianity in the porous system will be discussed in Chapter 7. The examination of heterogeneity in the system also provides a way to distinguish the mechanism of the enhancement.

For the development of the field of nanoionics, further research for the nanoscopic mechanism of the enhancement is expected and molecular dynamics simulation seems to be suitable for such treatments.

It is interesting to note that decreases in activation energy in the temperature dependence of conductivity have been reported for many systems showing enhanced dynamics.

In the case of fragile ionic liquid, non-Arrhenius behavior of the diffusivity is explained by the inflection of the behavior at $T_c$ (or $T_B$) [38] in diffusivity against the inverse of temperature on a double-logarithmic scale. Above it, cooperative motions with long-ranged motion are dominant, while at lower temperatures localized motion with shorter length scales becomes dominant. In other words, it is related to the existence of heterogeneous dynamics, where the long-ranged cooperative jumps are suppressed at lower temperatures. This is due to the formation of more closely packed structures [39].

From this point of view, observed behavior in composites (and porous systems as shown in Chapter 6) is regarded as the result caused by the weakened suppression of the motions at lower temperatures by introducing pores (free volume).

## 3.2 Nanostructures of Aggregates and Gels

Besides the enhancement of ion dynamics, spontaneous formation of the gel from the nanocolloidal solutions and its complex structures are the main problems treated in this book. Hereafter, the situation of the complicated structures of aggregates and gels known in experimental works will be summarized. Among nanoscale structures, colloidal systems play important roles in many fields, including the pharmaceutical and biotechnology fields. Both obtaining the well-dispersed colloidal solutions and controlling the development of aggregates and gels are targets of challenges in colloid chemistry. This is also important in industry and environmental chemistry. Specifically, extracting heavy metals from water coagulation using colloidal silica or sodium silicate is a useful technique. Therefore, it is an important task to learn the mechanism of coagulation. In this section, several examples of studies of the formation of gels found in the history of colloidal chemistry will be shown.

### 3.2.1 Fractal Dimension of Aggregates

Fractal dimension [40, 41] is useful to characterize the system with complex structures as already discussed in Section 2.2. The value can be connected to other properties.

For example, dimension ($D_f$) of the clusters formed during the aggregation of colloids are used to discuss the mechanism of formation [42]. In the work by Gauer and coworkers [43], diffusion-limited cluster-cluster aggregation (DCLA) process of tetrafluoroethylene was examined and the dimension seems to reflect correctly the degree of coalescence among the particles. Later, the fast small-angle light scattering (SALS) technique [44] has also been used to determine the $d_f$ value during the aggregation. It was found that in the diffusion-limited aggregation regime, the $d_f$ value can be correctly determined from both the power-law regime of the average structure factor of the clusters and the scaling of the zero-angle intensity versus the average radius of gyration.

Experimentally, Sorensen and coworkers [45] have observed the existence of two length scale regions in the structure of soot as shown in Fig. 3.7. It is called "super-aggregate," that is, aggregate of aggregates. Similar structures can be reproduced by off-lattice Monte Carlo methods of diffusion-limited (DLCA), ballistic limited, and reaction-limited cluster aggregation classes [46]. The result of DLCA in that work is shown in Fig. 3.8. In the case of light scattering for a system $N$ discrete scattering centers, intensity, $I(\mathbf{q})$, for elastically scattered light is represented as follows:

$$I(\mathbf{q}) = NF(\mathbf{q})S(\mathbf{q}),\ \mathbf{q} = |\mathbf{q}| = \frac{4\pi}{l}\sin\left(\frac{\theta}{2}\right), \tag{3.1}$$

where $S(\mathbf{q})$ is the static structure factor for wave number vector $\mathbf{q}$ and $l$ is the wavelength. When the radius of gyration, $R_g$, is defined by

$$R_g^2 = \frac{\int r^2 \rho(\mathbf{r}) d\mathbf{r}}{\int \rho(\mathbf{r}) d\mathbf{r}}, \tag{3.2}$$

The $R_g$ value for the onset of gelation is represented

$$R_{g,G} = a\left[k_0^{-1}\left(\frac{D_f+2}{D_f}\right)^{d/2} f_v\right]^{-1/d(d-D_f)}, \tag{3.3}$$

where $f_V$, is the monomer volume fraction of the entire system, which is assumed to be the same as the fraction in individual clusters.

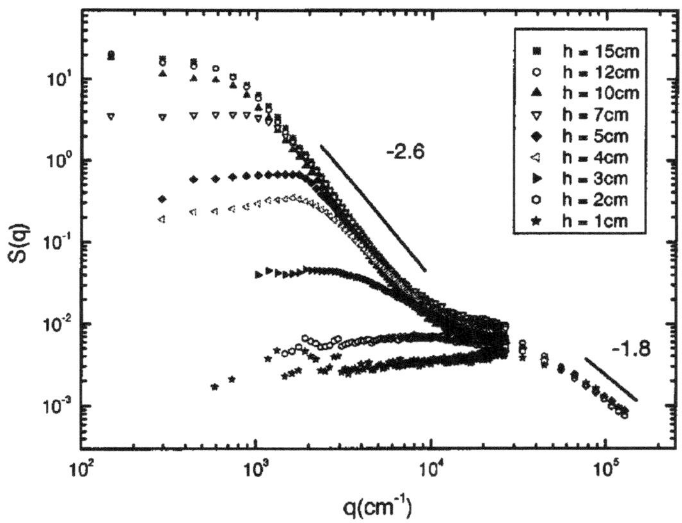


**Figure 3.7** Scattered light intensity as a function of the wave vector for an acetylene/air diffusion flame with a flow rate of 6.6 cm$^3$/s at different heights above the burner, h. Reprinted with permission from Sorensen, C. M., Kim, W., Fry, D., Shi, D., Chakrabarti, A., Observation of soot super-aggregates with a fractal dimension of 2.6 in laminar acetylene/air diffusion flames, *Langmuir,* 19, 7560–7563, (2003). Copyright (2003). Americal Chemical Society.

Experimentally, another technique is also used. Weitz and coworkers [47, 48] identified two distinct regions of aggregates with different fractal dimensions in experiments based on TEM (transmission electron microscopy) images. They observed fractal dimensions of approximately $D_f \sim 1.8$ for diffusion-limited colloid aggregation and $D_f \sim 2.1$ for reaction-limited colloid aggregation, and the results were considered rather universal. This approach applies to many systems and can be a powerful tool to understand the properties of aggregates and gels.

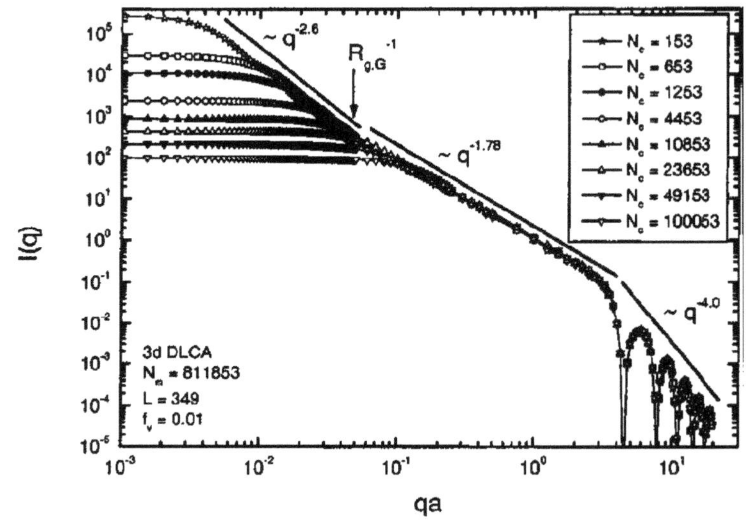


**Figure 3.8** Structure of the largest cluster is studied by computing $I(\mathbf{q})$ according to Eq. (3.1) at different times during aggregation for the three-dimensional DLCA simulation with a monomer volume fraction $f_v = 0.01$. Here, $N_c$ is the number of clusters at a given time during aggregation. Thus, smaller $N_c$ corresponds to a later time. The location of the calculated ideal gel point radius of gyration [Eq. (3.3)] is shown as $R_{g,G}^{-1}$. For very large **q**, probing the smallest length scale in the system, i.e., the monomer radius a, one observes "interference ripples" and Porod's law [$I(\mathbf{q}) \sim \mathbf{q}^{-4}$] in this log-log plot. For intermediate values of **q**, Isqd characteristic of a DLCA aggregate is seen with a fractal dimension of $D_f \cong 1.8$ obtained from the slope of the graph. Near the gel point, **Isqd** evolves to show two power-law regimes. The first regime mentioned above remains intact, but at smaller **q**, hence at a larger length scale, a $D_f \cong 2.6$ regime appears. For smaller values of **q**, the power law gives way to a **q**-independent Rayleigh regime. Solid lines have been drawn at the designated slope ($\sim \mathbf{q}_f^{-D}$) and serve only to guide the eye. Reprinted with permission from Fry, D., Chakrabatti, A., Kim, W., Sorensen, C. M. (2004), Structural crossover in dense irreversibly aggregating particulate systems, *Phys. Rev. E*, 69, 061401(1–10). Copyright (2004) by the American Physical Society.

To understand the results of these works, caution might be required for the difference among different methods. When a two-dimensional image is used for the analysis, the procedure of the projection to the 2-D plane modified the dimension of the system. If one applies both two- and three-dimensional methods to the same system, the problem becomes clearer. In Chapters 8 and 9, such a comparison will be done for the systems prepared by atomistic MD simulations [49]. It is worth mentioning that the existence of the two length scale region, that is the formation of "super-aggregate" discussed in refs. [45] and [46], is well reproduced by the full atomistic MD simulation of aggregates and the gel, without assuming any mechanism of formation, *a priori*.

Several methods and related concepts such as fractal dimensions [40, 41], $Q_n$ distributions [50, 51] are introduced to explain the complex structures of clusters, aggregates, and gels.

Because the concept of mono-fractal is not necessarily adequate to represent complex systems, the concept of multifractal will be introduced in Chapter 4. As shown in this chapter, one should not forget the importance of the cooperativity of motions of ions (particles, molecules) found in ionic conductors [1] as well as in other glass-forming systems [52]. One needs to examine how such motions are modified in more complex materials such as porous materials and/or composites. In Section 2.2, the classification of structures in colloidal systems is discussed based on their multifractality.

## 3.3 Conclusion

In this chapter, interesting problems in nanotechnology, especially in nanoionics and colloidal chemistry, have been summarized and some points to treat the systems have been clarified. Several factors that affect the enhancement of ionics have been examined. The problem seems to be still open to researchers. Some of the difficulties of solving problems lay on the conditions of the complex systems, which much depends on how to prepare samples. Furthermore, it is not easy to separate the factors affecting dynamics in experiments. In this situation, the problem may be expected to be studied by molecular dynamics simulations with well-controlled conditions. The enhanced dynamics are affected by macroscopic structures such as layers, distribution of particle sizes, grain boundaries as well as microscopic structures, and impurities. In some explanations, alumina or silica are just treated as insulators. In Chapter 6, it will be shown that introducing pores into the systems affects diffusivity (hence conductivity) of the systems, where caged ion dynamics and jump paths are strongly affected by them. Another important problem treated in this book is colloidal systems, which show the spontaneous formation of gel.

As shown in this chapter, one of the useful methods to characterize the complex system such as gels is the fractal dimension analysis, and the existence of several length scale regions means the multifractality of the structures. These facts should be considered for the further development of these fields.

## References

1. Habasaki, J., León, C., Ngai, K. L. (2017), *Dynamics of Glassy, Crystalline and Liquid Ionic Conductors, Experiments, Theories, Simulations*, Springer International, Switzerland.
2. Liang, C. C. (1973), Conduction characteristics of the lithium iodide-aluminum oxide solid electrolytes, *J. Electrochem. Soc.*, 120(10), pp. 1289–1292; doi:10.1149/1.2403248.
3. Ardel, G., Golodnitsky, D., Peled, E., Wang, Y., Wang, G., Bajue, S., Greenbaum S. (1998), Bulk and interfacial ionic conduction in $LiI/Al_2O_3$ mixtures, *Solid State Ionics*, 113–115, pp. 477–485.
4. Stoneham, A. M., Wade, E., Kilner, J. A. (1979), A model for the fast ionic diffusion in alumina-doped LiI, *Mater. Res. Bull.*, 14, pp. 661–666.
5. Wagner, C. (1972), The electrical conductivity of semi-conductors involving inclusions of another phase, *J. Phys. Chem. Solids*, 33, pp. 1051–1059.
6. Maier, J. (1985), Space charge regions in solid two-phase systems and their conduction contribution—I. Conductance enhancement in the system ionic conductor-'inert' phase and application on AgCl:$Al_2O_3$ and AgCl:$SiO_2$, *J. Phys. Chem. Solids*, 46, pp. 309–320.
7. Maier, J. (1995), Ionic conduction in space charge regions, *Prog. Solid State Chem.*, 23, pp. 171–263.

8. Knauth, P. (2000), Ionic conductor composites: Theory and materials, *J. Electroceramics,* 5, pp. 111–125. https://doi.org/10.1023/A:1009906101421.
9. Dudney, N. J. (1988), Enhanced ionic conductivity in composite electrolytes, *Solid State Ionics,* 28–30, pp. 1065–1072.
10. Dudney, N. J. (1987), Enhanced ionic conduction in AgCl-$Al_2O_3$ composites induced by plastic deformation, *J. Am. Ceram Soc.*, 70(2), pp. 65–68.
11. Dudney, N. J. (1985), Effect of interfacial space-charge polarization on the ionic conductivity of composite electrolytes, *J. Am. Ceram Soc.*, 68, pp. 538–545.
12. Skarstad, P. M., Merritt, D. R., Owens, B. B. (1981), Conductivity enhancement in lithium iodide monohydrate by dispersed oxide phases, *Solid State Ionics,* 3, pp. 277–281.
13. Bjorkstam, J. L., Villa, M., Rahman, M., Skarstad, P. M. (1983), Lithium-ion diffusion and water molecule reorientations in lithium iodide monohydrate, *Solid State Ionics*, 9–10, pp. 111–118.
14. Chen, L. Huang, X., Kelder, E., Schoonman, J. (1995), Diffusion enhancement in $Li_xMn_2O_4$, Solid State Ionics, 76, pp. 91–96.
15. Roman, H. E., Bunde, A., Dieterich, W. (1986), A percolation model with two critical points, *Phys. Rev. B*, 34, pp. 3439–3445.
16. Dieterich, W., Dürr, O., Pendzig, P., Bunde, A. (1999), Nitzan, Percolation concepts in solid state ionics, *Phys. A*, 266, pp. 229–237.
17. Landauer, R. (1952), The electrical resistance of binary metallic mixtures, *J. Appl. Phys.,* 23, pp. 779–784.
18. Heitjans, P., Wilkening, M. (2009), Ion dynamics at interfaces: Nuclear magnetic resonance studies, *MRS Bull.*, 34, pp. 915–922, www.mrs.org/bulletin 915–922.
19. Schreck, E., Läuger, K., Dransfeld, K. (1986), Enhanced ionic conductivity at the interface between sapphire and solid lithium iodide, *Z. Phys. B*, 62, pp. 331–334.
20. Tadanaga, K., Imai, K., Tatsumisago, M., Minami, T. (2002), Conductivity enhancement and thermal properties of AgI-$MO_2$ (M = Zr, Si) composites using sol-gel derived aerogels, *J. Electrochem. Soc.,* 149, pp. A773–A777.
21. Ruprecht, B., Wilkening, M., Steuernagel, S., Heitjans, P. (2008), High anion conductivity in a ternary non-equilibrium phase of $BaF_2$ and $CaF_2$ with mixed cations, *J. Mater. Chem.,* 18, pp. 5412–5416.

22. Phipps, J. B., Whitmore, D. H. (1983), Ion transport in LiI-$SiO_2$ composites, *Solid State Ionics*, 9/10, pp. 123–130.

23. Yamada, H., Moriguchi, I., Kudo, T. (2005), Nano-structured Li-ionic conductive composite solid electrolyte synthesized by using mesoporous $SiO_2$, *Solid State Ionics*, 176, pp. 945–953.

24. Saha, D. R., Nandi, A. K., Chakravorty, D. (2013), Enhancement of ionic conductivity in $Li_2O$–$SiO_2$ glass in nanodimensions grown within pellets of ZnO nanorods and magnetodielectric properties of these nanocomposites, *J. Non-Cryst. Solids*, 376, pp. 12–17.

25. Borodin, O., Zhuang, G. V., Ross, P. N., Xu, K. (2013), Molecular dynamics simulations and experimental study of lithium ion transport in dilithium ethylene dicarbonate, *J. Phys. Chem. C*, 117(15), pp 7433–7444.

26. Habasaki, J., Okada, I., Hiwatari, Y. (1997), Fracton excitation and Lévy flight dynamics in silicate glasses, *Phys. Rev. B*, 55, pp. 6309–6315.

27. Lin, D., W. Liu, L. Y., Lee, H. R., Hsu, P.-C., Liu, K., Cui, Y. (2016), High ionic conductivity of composite solid polymer electrolyte via in situ synthesis of monodispersed $SiO_2$ nanospheres in poly(ethylene oxide), *Nano Lett.*, 16, pp. 459–465.

28. Habasaki, J., Ngai, K. L. (2015), Molecular dynamics study of heterogeneous dynamics in lithium disilicate crystal, *J. Electroceram*, 34, pp. 43–56.

29. Küchler, R., Kanert, O., Vereget, T., Jain, H. (2007), Effect of devitrification on ion motion in lithium-disilicate glass, *J. Non-Cryst. Solids*, 353, pp. 3940–3946.

30. Glass, A. M., Nassau, K., Negran, T. J. (1978), Ionic conductivity of quenched alkali niobate and tantalate glasses, *J. Appl. Phys.*, 49(9), 4808–4811.

31. Klafter, J., Shlesinger, M. F., Zumofen, G. (1996), Beyond brownian motion, *Phys. Today*, 49(2), pp. 33–39, and references therein.

32. Deng, Y. Eames, C., Chotard, J.-N., Islam, M. S. (2015), Structural and mechanistic insights into fast lithium-ion conduction in $Li_4SiO_4$-$Li_3PO_4$, solid electrolytes, *J. Am. Chem. Soc.*, 137(28), pp. 9136–9145. DOI: 10.1021/jacs.5b04444.

33. Panchmatia, P. M., Armstrong, A. R., Bruce, P. G., Islam, M. S. (2014), Oxygen diffusion in yttria stabilised zirconia—experimental results and molecular dynamics calculations, *Phys. Chem. Chem. Phys.*, 16, pp. 21114–21118.

34. Nishino, S., Fujiwara, T., Yamasaki, H. (2014), Nanosecond quantum molecular dynamics simulations of the lithium superionic conductor $Li_{4-x}Ge_{1-x}P_xS_4$, *Phys. Rev. B,* 90, 024303 (1–8).

35. Fujimura, K., Seko, A., Koyama, Y., Kuwabara, A., Kishida, I., Shitara, K., Fisher, C. A. J., Moriwake, H., Tanaka, I. (2013), Accelerated materials design of lithium superionic conductors based on first-principles calculations and machine learning algorithms, *Adv. Energy Mater.*, 3, pp. 980–985.

36. Habasaki, J., Hiwatari, Y. (2002), Molecular dynamics study of dynamical heterogeneity in ion conducting glasses, *J. Therm. Anal. Calorim.*, 69, pp. 1005–1013.

37. Habasaki, J., Ngai, K. L., Hiwatari, Y. (2002), Molecular dynamics study of cage decay, near constant loss, and crossover to cooperative ion hopping in lithium metasilicate, *Phys. Rev. E,* 66, 021205 (1–11).

38. Habasaki, J., Ngai, K. L. (2008), Molecular dynamics study of the dynamics near the glass transition in ionic liquids, *Anal. Sci.,* 24, pp. 1321–1327.

39. Habasaki, J., Ngai, K. L. (2015), Rigidity and soft percolation in the glass transition of an atomistic model of ionic liquid, 1-ethyl-3-methyl imidazolium nitrate, from molecular dynamics simulations—existence of infinite overlapping networks in a fragile ionic liquid, *J. Chem. Phys.,* 142, 164501(1–19).

40. Ben-Avraham, D., Havlin, S. (2000), *Diffusion and Reactions in Fractals and Disordered Systems,* Cambridge University Press, Cambridge.

41. Mandelbrot, B. (1967), How long is the coast of Britain? statistical self-similarity and fractional dimension, *Science,* 156, pp. 636–637.

42. Meakin, P. (1987), Fractal aggregates, *Adv. Colloid Interface Sci.,* 28, pp. 249–331.

43. Gauer, C., Wu, H., Morbidelli, M. (2009), Control of coalescence in clusters of elastomer colloids through manipulation of polymer composition, *Macromolecules,* 42 (22), pp. 9103–9110.

44. Wei, D., Wu, H., Xia, Z., Xie, D., Zhong, L., Morbidelli, M. (2012), Monitoring coalescence behavior of soft colloidal particles in water by small-angle light scattering, *Colloid Polym. Sci.,* 290, pp. 1033–1040.

45. Sorensen, C. M., Kim, W., Fry, D., Shi, D., Chakrabarti, A. (2003), Observation of soot superaggregates with a fractal dimension

of 2.6 in laminar acetylene/air diffusion flames, *Langmuir,* 19, pp. 7560–7563.

46. Fry, D., Chakrabatti, A., Kim, W., Sorensen, C. M. (2004), Structural crossover in dense irreversibly aggregating particulate systems, *Phys. Rev. E,* 69, 061401(1–10).
47. Weitz, D. A., Huang, J. S., Lin, M. Y., Sung, J. (1985), Limits of the fractal dimension for irreversible kinetic aggregation of gold colloids, *Phys. Rev. Lett.*, 54, pp. 1416–1419.
48. Lin, M. Y., Lindsay, H. M., Weitz, D. A., Ball, R. C., Klein, R., Meakin, P. (1989), Universality in colloid aggregation, *Nature*, 339, pp. 360–362.
49. Habasaki, J., Ishikawa, M. (2014), Molecular dynamics study of coagulation in silica-nanocolloid–water–NaCl systems based on the atomistic model, *Phys. Chem. Chem. Phys.*, 16, pp. 24000–24017.
50. Greaves, G. N., Sen, S. (2007), Inorganic glasses, glass-forming liquids and amorphizing solids, *Adv. Phys.*, 56, pp. 1–166.
51. Habasaki, J., Ngai, K. L. (2013), Molecular dynamics study of network statistics in lithium disilicate: Qn distribution and the pressure-volume diagram, *J. Chem. Phys.*, 139, 064503 (1–14).
52. Ngai, K. L. (2011), *Relaxation and Diffusion in Complex Systems*, Springer, New York.

Chapter 4

# Fundamentals of Molecular Dynamics (MD) Simulations and Tools for Examining Nanostructured Materials

In chemistry, physics, biochemistry, and materials sciences, computer simulations are useful methods to examine the nanostructures and dynamics in nanoscopic levels. Molecular dynamics simulations (MD) are one of the principal methods of computational science. In this method, positions and trajectories of particles (atoms, molecules) are determined by solving the equation of motion within a computer. In this book, our discussion is mainly focused on classical MD [1–3] using the potential parameters based on the *ab initio* calculations. Nowadays, *ab initio* MD simulations such as the Car–Parrinello method [4, 5] using density functional theory have become important. However, the application of them for complex materials with covering the suitable length and time scales is not necessarily easy, especially in the *ab initio* method, which requires more resources and times for calculations. Common problems for these methods will be discussed.

In this chapter, after a short description of the fundamentals of MD simulations, several functions and tools useful for the

*Molecular Dynamics of Nanostructures and Nanoionics: Simulations in Complex Systems*
Junko Habasaki

ISBN 978-981-4800-77-8 (Hardcover), 978-1-003-04490-1 (eBook)
www.jennystanford.com

characterization of the ionic system as well as nanostructured materials are introduced with examples of a molten NaCl system. From these functions, one can determine several length scales relevant to ionic systems.

Fractal dimension analyses are useful tools to examine the complexity of structures and dynamics. Then results of fractal dimension analyses of density profiles used for characterization of structures in lithium silicates and ionic liquid (IL), (1-ethyl 3-methyl imidazolium nitrate, EMIM-$NO_3$) will be shown. Dynamics of these systems are explained by the multifractal nature of the profile and the existence of corresponding two-length scale regions of trajectories of walks. Namely, through multifractal analyses of density profiles and walks, structures can be connected to dynamics. This kind of analysis provides a common framework for the characterization of more complex systems.

Nanostructured systems often consist of networks of constituents and hence, other approaches introduced here are analyses of network structures found in silicates and related systems. Several tools to characterize networks will be explained.

Of course, some of these methods are applicable to simulations by other methods as well as experimental results.

## 4.1 Purpose of MD Simulations

Comparison of results of experiments and MD is an important task; however, the purposes of the MD simulations are not necessarily just reproducing systems. Some of the possible purposes of MD simulations (or other simulations) are classified as follows:

1. They can be used to reproduce structures and properties to mimic a realistic system for comparison with experiments and further examination of details of structures and dynamics. It is also useful to examine the morphology of nanostructures and mechanisms underlying them. Of course, careful judgment is required if it can be comparable or substitute for other experimental methods.
2. They can be used to examine essential physics or chemistry underlying the behaviors of the system both at microscopic

and macroscopic levels. Transport properties, structures, and mechanical properties under several situations with modified shapes and/or sizes in different states can be examined. One can examine what happens if the potential models are modified or coarse-grained. Adequate modeling of the system by MD simulations itself is useful to understand the characteristics of the systems.

3. It may be worth mentioning that simulations are highly safe and have a low environmental burden. They can be used to examine the system, which is difficult to treat in experimental methods due to its toxicity, explosive nature, deliquescence, etc.
4. They can be used to examine the system under extreme conditions, such as high pressure and high temperature, which is difficult to treat in experiments.
5. They can be used to design new materials with modifying components, mass, size of particles, etc., or considering new combinations.
6. They can be used for the screening of a suitable system and/or material to fit the usage of them. For example, they are used for a survey of candidates for pharmaceuticals.
7. If the quality of the potential model (for both parameters and functional forms) used is good enough, the model has a power of prediction. Therefore, simulations can be used for the predictions of structures and properties of materials including new materials.
8. They can be used to examine the effects of modification and/or of introducing the complexity of the systems, such as the mixing of components.
9. Simulations are also used to examine the validity of concepts and/or theories as well. If the existent theory or model is not adequate for each purpose, simulations can add useful knowledge, suggestions, answers, and/or new questions.

These purposes are not necessarily separable. For example, a modification of the transport properties of ions by porosity treated in this book is related to 1, 2, 7, and 8. Also, they are not limited to the list mentioned above.

Thus, MD simulations are useful for both applications and more fundamental understanding of physical and/or chemical concepts.

One of the systems covered in this book is the nanocolloidal silica-water-salt system in fully atomistic simulations, although only the coarse-grained level ones tend to be used in colloidal science. The latter treatment is probably coming from the large size difference between colloids and solvent molecule; however, such coarse-gaining is not necessarily suitable for the treatment of nanostructures and dynamics related to them. Since both system size and time scale required to simulate such systems are quite large, it is challenging to do atomistic type simulations.

Once details of atomistic or molecular level understandings are obtained enough, some coarse-grained approaches are also useful. Sometimes simulations of different fining levels are connected to form more macroscopic systems.

## 4.2 Treatment of Complex Structures

In many applications, systems are mixtures of several components and/or the part of more complicated systems. Further treatment of the system having hierarchy including several materials will be necessary.

There are different approaches to treat complex structures. Here these situations are classified as follows for typical situations:

1. Direct simulations, including all components and substructures of the system in the simulations. If it is possible, it includes the interaction among components and hierarchy structures without assumptions for the rules of the formation of structures.
2. The situation is like 1 but each component is substituted by more coarse-grained level ones.
3. Each component is formed in the atomistic-level simulations (in both classical and *ab initio* methods can be used here). These components are combined with some rules. If each structure has a simple role as a module of electrical circuits such as resistance, condenser, etc., this kind of approach is possible.
4. Several components form higher levels of substructures.

For applications, simulations of more coarse-grained levels might be necessary in several cases. Combinations of methods with several different levels of coarse-graining can be used.

The selection of the methods depends on the purpose of the simulations and may be different case by case. The following consideration appears to be useful for choosing a suitable method.

The coarse-grained systems are useful in the following situations. If feedback from other components is correctly included, the coarse-grained level approach will be successful. When the time scale of the event within a component is considerably different from the event among components, it can avoid the undesired coupling among components. In these cases, key features to select the suitable approaches depend on the magnitude of the coupling or interaction among substructures or domains.

Let's consider the following examples to determine what levels of simulations are suitable. First, the situation to treat ionic motion is considered. When we consider fast ionic conducting systems, the motion of fast ions is known to be de-coupled with the motion of networks forming the ion channels. However, cages of the ions must be flexible for higher ionic conductive systems, and therefore it is difficult to separate or coarse-grained the network part and ions to represent the behavior of ions.

That is, for ionic motions, atomistic-level simulations seem to be required to treat the different kinds of couplings between ions and the matrix.

The treatment with different levels of coarse-graining seems to be useful for some cases to treat ILs [6], which have a more complex inner structure of ions. Ions can be treated as large ions without structures or ions with simplified inner structures.

Another example is the gelation of colloidal systems discussed in Chapters 8 and 9. During atomistic simulations, structures of water are found to be changed. This means that approximation by continuum solvents is not good enough to represent such situations.

If there are no interactions among different levels of hierarchy structures, simulation systems with switching the different levels can be used for complex systems.

## 4.3 Molecular Dynamics Methods Used for Material Designs

In this book, we focus on the classical molecular dynamics simulations using potential parameters based on the *ab initio* calculations, although sometimes Monte Carlo simulations and *ab initio* MD simulations such as Car–Parrinello methods and combinations of these methods are referred to. The merit of the first approach is the applicability of relatively larger systems, long time scales with keeping the characteristics of essential properties obtained by *ab initio* methods. Due to these characteristics, methods can be used for the designing of materials for more complex systems with hierarchy and with more components. One major problem of this approach is that one may encounter the situation of a missing parameter, with an increasing number of components and their combinations. It may be necessary to develop new parameter sets suitable for each purpose. Some parameter sets with considering transferability and/or consistency might be found in the literature. Even in this case, it is desirable to check the applicability and suitability in light of the purpose.

## 4.4 Solving the Equation of Motion

In principle, MD simulations describe the motions of particles (ions or atoms) by solving the equation of the motion. That is, the position of the particle is predicted from the previous and present ones by adding the forces acting from other particles. In the case of classical MD, it would not be an exaggeration to say that potential functions and its parameters (force fields) determine the fate of the particles, if other conditions are reasonably selected.

The equation of the motion for the *i*-th particle can be written as

$$\frac{d^2\mathbf{r}_i}{dt^2} = \frac{\mathbf{F}_i}{m}, \quad i = 1,2,\ldots N. \tag{4.1}$$

### 4.4.1 Algorithm

In classical MD simulations, the equation of the motion is solved numerically. In other words, the next position of the *i*-th particle is calculated based on the position at $t$ and that of one or several steps before. Several algorithms to solve the equation of motions have so far been proposed. Verlet algorithm [7] and Gear's method [8], which is one of the predictor-corrector methods with several steps, are also used for MD simulation [1–3].

The former is suitable for the calculation of motion including jump or hops of ions often encountered in the simulations of melts and glasses because it does not drag the memory of past motions for a long time.

Verlet method is explained as follows [7]:

$$\mathbf{r}_i(t+\Delta t) = \mathbf{r}_i(t) + \Delta t\,\dot{\mathbf{r}}_i(t) + \frac{(\Delta t)^2}{2}\frac{\mathbf{F}_i(t)}{m} + O((\Delta t)^3) \tag{4.2}$$

$$\mathbf{r}_i(t-\Delta t) = \mathbf{r}_i(t) - \Delta t\,\dot{\mathbf{r}}_i(t) + \frac{(\Delta t)^2}{2}\frac{\mathbf{F}_i(t)}{m} + O((\Delta t)^3), \tag{4.3}$$

where the vector $\mathbf{r}(t + \Delta t)$ is for positions of the $i$-th particle after $\Delta t$ and $\mathbf{r}(t - \Delta t)$ is that before $\Delta t$.

One can obtain the following relation using the sum of Eqs. (4.2) and (4.3):

$$\mathbf{r}_i(t+\Delta t) + \mathbf{r}_i(t-\Delta t) = 2\mathbf{r}_i(t) + (\Delta t)^2\frac{\mathbf{F}_i(t)}{m} + O((\Delta t)^4), \tag{4.4}$$

while a following equation was obtained, from the difference of Eqs. (4.2) and (4.3):

$$\mathbf{r}_i(t+\Delta t) - \mathbf{r}_i(t-\Delta t) = 2\Delta t\,\dot{\mathbf{r}}_i(t) + O((\Delta t)^3), \tag{4.5}$$

That is, new position of the particle $i$ is represented by

$$\mathbf{r}_i(t+\Delta t) = 2\mathbf{r}_i(t) - \mathbf{r}_i(t-\Delta t) + (\Delta t)^2\frac{\mathbf{F}_i(t)}{m} + O((\Delta t)^4), \tag{4.6}$$

and a new velocity is represented by

$$\dot{\mathbf{r}}_i(t) = \frac{1}{2\Delta t}\{\mathbf{r}_i(t+\Delta t) - \mathbf{r}_i(t-\Delta t)\} + O((\Delta t)^2), \tag{4.7}$$

A numerical error in solving the equation of motion thus depends on the time of each step, $\Delta t$, and hence too large step should not be used. However, if the particle is localized for a long time, often a longer time step can be used because the structures of cages and or paths can have longer relaxation times. In many of ionic systems, each time step $\Delta t$ can be 1–4 fs. For the case of the system including light particles such as hydrogen, the smaller time step is necessary (typically, that less than 1 fs for the case of classical MD and even a shorter time is necessary for *ab initio* MD).

## 4.5 Required Conditions for MD Simulations

In conventional MD, periodic boundary conditions (PBC) are used to mimic the situation of bulk. When one uses PBC, caution is required for the size of the basic cell to be used. It may modify both the motion and the structure by repeating those in the basic cells. It also determines the wavenumber, which can be addressed.

Needless to say, good quality of potential parameters suitable for the purpose of the work is also required.

When one handles nanoscale structures with different treatments of boundary conditions, additional caution will be required. For example, if the surface of the materials is treated, often the region used for summation to take the statistical treatment needs to be changed and additional normalization procedures should be used for comparison with other systems.

In this chapter, some of these problems are discussed one by one.

### 4.5.1 Periodic Boundary Conditions and Finite Size Effects

To treat the properties of bulk, periodic boundary condition (PBC) is often used to remove the effects of the surface in MD simulations [2–4].

That is, the basic cell containing particles is surrounded by the periodically repeating image cells.

For a particle within the basic cell, interactions from particles within the sphere with a certain cutoff length (typically chosen to be $L/2$, where $L$ is a length of a basic cell) are considered, including ghost particles (images of the same particle found in the basic cell) in surrounding image cells. By this treatment, properties of the bulk can be simulated by the limited number of particles. If the particle moved beyond the boundary, the ghost particle comes into the basic cell. Therefore, in the case of constant number simulations such as *NVE* (constant number of particles, volume, and energy) or *NPT* (constant number of particles, pressure, and temperature) ensembles, the number of the particles is kept constant by the PBC.

Usually, Coulombic force can be calculated by the Ewald method [9] or by other methods such as particle mesh Ewald [10]. Even in the case of liquid or glasses can be treated by using the PBC, which mimics the periodicity like crystals. In many cases, the minimum image (within $2/L$) is used for the calculation of short-range forces. The same length is often used for the real part of the Ewald summation for the calculations of Coulombic forces.

In liquids or glasses, the structure shown by pair correlation function $g(r)$, decays within a certain length. For ion–ion interactions, the correlation of the structure (deviation from the line of $g(r) = 1$) is not clear at distance longer than 8–12 Å in ionic systems such as lithium silicate glass [3]. Therefore, about twice this length is a minimum length of the unit cell for MD simulations to avoid the ionic structure being affected by the PBC.

Sometimes, system size used in molecular dynamics simulations is not large enough and it causes several kinds of finite size effects in the results of simulations. This problem limits the usage of *ab initio* MD, which requires more calculation resources than classical MD. Even in the classical MD, it is not easy to treat the large system size required for the treatment of many systems with different purposes. Here we consider several origins of finite size effects to be considered.

1. Even when we considered the infinitive system using PBC, the wavenumber accessible by the simulation is limited by the size of the simulation box. Furthermore, when one considered the motion of particles in the system with PBC, it is repeated as well and some artificial waves or vibrations in the particle motions will be formed.
2. Especially in the case of crystals, a basic box of the MD is formed by several repeating basic lattices of the crystal

and therefore the number of repetitions of them in each axis direction will affect the periodicity of the motions. Consistently, mobility of ions is found to be affected by the system size considerably [11]. To reduce such effects, the system size used should be large enough to the possible extent, while ensuring the practical usability.

3. In the case of melt and glasses, cooperative motions of network formers and ions also may be affected by PBC. If the system size is too small, the particle might be affected by its own ghost in an image cell, which is moving in the same direction.

It is useful to change the box size to check the finite system size effect mentioned above. Interestingly, the size effect in nanosystems is also found in the study of ferroelectricity, characterized by a spontaneous electric polarization. Shimada and collaborators [12] have examined the ferroelectricity in edged $PbTiO_3$ nanowires under axial tension by *ab initio* density-functional theory calculations. They have reported that "surprisingly, the smallest PbO-terminated nanowire with a cross section of only one-unit cell can possess ferroelectricity while ferroelectricity disappears in the $TiO_2$-terminated nanowires with a cross section smaller than four-by-four cells diameter of about 17 Å. However, ferroelectricity is recovered by axial tension, where smaller nanowires require larger critical strains." These results are interesting to consider both the size effects and roles of the epitaxial strain.

### 4.5.2 System Size Effects Related to Crystallization

The remarkable size effect is also found in the calculation of thermal conductivity [13]. For the study of glasses, further caution is required to avoid the crystallization of the system. If only a small number of particles are contained in each basic box of the simulation, the system may easily crystallize and/or behaves like crystals because of PBC. When systems have network structures, long-ranged oscillation tends to be formed and continued by PBC. The motions of particles to be examined will be modified by these effects.

### 4.5.3 Required System Sizes and Related Conditions

Here we will discuss how large system size is required for meaningful arguments of structures and dynamics obtained by MD simulations. Such requirements are common for both classical and *ab initio* MD simulations. A part of arguments is found in [4] and is expanded for complex systems treated in the present book here. Of course, it will depend on the purpose of simulations, but here we consider some typical scales for the examination of structures and motions of particles such as the diffusion.

In general, one needs larger systems to access the structures and dynamics related to the smaller wavenumber. If the system size is too small, the particle might be affected by its own ghost in an image cell, which is moving in the same direction. In the case of slow dynamics near the glass transition temperatures or in viscose liquids, the size of the cooperatively rearranging region (CRR) of the systems [14, 15] should also be considered. With the changes in the size of CRR with temperatures and/or pressure of the system, the required system size will also change. Typical systems treated in recent MD works of ionic systems seem to be containing several thousand particles. However, a larger system size will be required if one examines the more complex systems with hierarchy structures and/or different domain sizes. The reason for the requirement of larger size can be easily found in the example of the structure of gels (see Chapters 2, 8 and 9.)

#### 4.5.3.1 Length scales in ionic systems such as Debye length

More specifically, we will consider some typical systems.

In simple ionic systems such as molten NaCl, long correlation length, including the Debye length, is found by the existence of charge density wave as discussed later in Section 4.8. Even for the crystals, larger systems will be required if one considers the fluctuation of the structures or long-ranged motions. Therefore, system size must be determined to represent these length scales reasonably.

Then we will consider the example of silicate glasses. In the strong system with three-dimensional networks, a large finite-size effect is known, while the effect is relatively small in fragile liquids [16]. If attention is paid to the local structures in a fragile system with larger alkali contents, relatively small systems can be used, while if one would like to discuss the network statistics such as the distribution of rings, larger system size will be necessary. Of course, it is better to check several length scales in the systems. As discussed in Chapter 2, these length scales can be modified by nano-sizing. They will be modified by other perturbations such as the mixing of components and formation of composites.

Next, we will consider the example of simple ionic liquid ($EMIM^+$-$NO_3^-$), (here $EMIM^+$ means 1-methyl 3-ethyl imidazolium ion). In this system, correlation in $g(r)$ is observed until ~18 Å at 400 K and therefore the system size with ~36 Å of $L$ will be a good choice for many purposes, while larger cell size will be required if the structure has long tails (chains) and/or if one is considering the existence of large-scale domains.

Besides the system size, other conditions such as the shape of the basic cell may need to set in the MD simulations, especially for complex systems. When some alignments of molecules or ions are found, a non-cubic basic box of the simulations might be better to use.

In the study of nanostructured systems such as a nanoparticle and/or each isolated system, the system without PBC might be used. In that case, a direct sum or by multipolar expansion can be used for calculations of the Coulombic Forces. If one uses the system without PBC, additional care is necessary for the calculation of the structural properties such as pair correlation functions $g(r)$. Usually, the function is calculated using forces within a radial distance $r$ and normalized by the number of the particles within the sphere. It needs to be modified by the structures of boundary, dimensionality, and shapes in the case of nanostructured systems.

As shown in Chapters 8 and 9 for the study of aggregates and gels, the contribution of large length scales is found. For example, to treat nanocolloidal-silica–water–NaCl systems in Chapters 8 and 9, the system containing tens of thousands of particles was used. Even a larger system is required in some objectives of works. A similar situation is also observed in other systems, such as network-forming systems and/or composites.

With an increase in systems size, the time required for equilibration of the system also becomes longer. For the simulation of equilibrated properties of large systems, one should be careful to equilibrate systems sufficiently, although, for some special purposes related to the problem of the glass transition, rapidly quenched systems are also examined. The required equilibration time is much affected by the dynamics of the particles in the system. In the system with slower dynamics, longer equilibration time is required.

### 4.5.4 Time Scales to Be Covered

In non-equilibrium systems such as supercooled liquids and glasses, non-equilibrium relaxation, that is aging, overlaps. In such a case, temperature increases gradually during the runs in the *NVE* condition and this fact is useful to distinguish the status of the system. Due to this reason, we need to distinguish the term such as "*NVE* condition" from "*NVE* ensemble" in some works [17]. For example, in the case of the porous system formed in *NVE* conditions, relaxation of the system to form the larger voids is found as shown in Chapter 6. This formation of larger voids caused by the stress relaxation is closely related to the mechanism of the fracture. If one examines the mechanical response of the system, the role of stress relaxation time and its relation to the conditions of MD should be considered. It should be noted that the fracture in *NPT* conditions in MD simulations is not necessarily directly comparable to the experiments under atmospheric conditions. Crystallization also overlaps the equilibration process.

In the case of *ab initio* MD simulations, available equilibration time may be restricted, and hence, using the equilibrated structure obtained by classical MD will be a good choice for the initial configurations, if available.

Due to the complexity of structures, dynamics of ions as well as atoms, molecules in nanostructured materials may contain several different time scales of dynamics. Even simple glass-forming materials and/or simple ionic system, it shows several length-scale and time-scale regions in the mean squared displacement (MSD), which corresponds to the caging to diffusive regimes through the power law region (see Chapter 5). Sometimes, a time region of local motions seems to be erroneously regarded as the diffusive regime, and hence, a careful judgment of the regions to be examined is required.

### 4.5.5 Relationship between Surfaces of Nanostructured Materials and Periodic Boundary Conditions in MD Simulations

In the field of molecular dynamics (MD) simulations, PBC is usually used to reduce the effects of the surfaces when one treats the bulk systems as already mentioned.

However, even though such conditions to mimic the infinite structure are applied, it is known that properties of systems strongly depend on the size of a basic box used in the simulations and the number of particles included there. Such finite size effects are caused by restriction by introduced repeating structures and motions imposed by PBC as already discussed. That is, the length scale, which can be treated in the simulation, is limited by the system size. This fact is interesting to consider the situation of nanostructured materials.

The situation of some of the nanostructured materials such as nanoparticles is comparable to the MD cell for which PBC is removed. Some systems may not be stable under such conditions. Obviously, in this case, long-ranged structure and motions are terminated at the surface. In other words, nanoparticles or films, materials trapped in pores, properties are affected by the absence of the short and long length-scaled structures and cooperative motions at the surface. Of course, rearrangement of atoms occurred on the surface.

This situation will be enhanced in further nano-sizing.

It is possible to simulate a surface of nanoparticles without PBC. Actually, simulations of small clusters have been done without PBC in several works. For example, in the simulation of several sizes of the cluster of silica [18], it was found that smaller clusters have a higher density, but the values are in close agreement with that calculated for bulk silica in the coesite phase. If PBC is removed to mimic the surface of the nanoparticles, one should not forget that the long scale forces and dynamics are also cut by it. That is, properties of nano-sized particles are not necessarily caused by the effects of the large surface area alone but the inclusion of the changes in structures and their length scales.

As shown in Chapter 8, nanocolloidal silica was prepared as the cluster of $SiO_4$ units without PBC [19] in our MD simulations. Then such model colloids are used as a part of the colloidal solutions.

## 4.6 Calculation of Coulombic Forces

For the simulations of ionics, the treatment of the Coulombic force of long range is included. Traditionally, Ewald methods [7] and/or related methods such as particle mesh Ewald [8] are used for long-range Coulombic force. Multipolar expansion [20] is also applicable even when the system contains defects. Recently, Wolf method [21, 22] is also used in ionic systems in which the Coulomb force is truncated at a certain length because the force was considered not necessarily long scale due to the screening by other ions. The method is useful to save the cost of the calculation in several situations in ionic systems [23]. Charge neutrality of the system within a basic cell of MD, which is usually imposed to the system, allows to use the Wolf method in disordered system [23]; however, the charge neutrality imposed to the cell (equal number of cation and anion in the case of monovalent systems.) itself is one of the causes of the finite-size effects of the MD simulations, since it causes at least some structural constraints. Therefore, a careful choice of the system size is also required. The check of the applicability of each method in each case seems to be the responsibility of researchers. In the case of molten salt and/or ionic liquids as well as network systems and more complex systems, the Coulombic force is not necessarily screened within a short length scale and such length scale is a strong function of temperature and/or compositions as shown later. In the following sections, the existence of several length scales in some ionic systems including IL will be shown.

In nanostructured materials without PBC, the Ewald method and related one assuming periodicity cannot be used. A careful choice of the method of calculation of the Coulombic force is necessary.

## 4.7 Treatment and Tools of Nanostructured Materials in MD Simulations

MD simulations often treat the nanoscaled systems with and without periodic boundary conditions, and therefore general structural analyses mentioned in the following sections can be applied to nanostructured materials. Tools explained here are pair correlation functions, running coordination numbers, from which one can determine several length scales and fractal dimensions. Fractal dimension analyses [24, 25] are expanded to multifractal structures [26–28] as well as those for dynamics. Analyses of network structures, by using $Q_n$ distributions [29–31], cluster analyses using a dendrogram [32] are also introduced and some of them will be applied to data obtained by MD simulations in Chapter 9.

In nanoparticles and/or porous systems, the roles and characteristics of the surface area are often targets of debates. Porosity and connectivity of pores are also used to characterize porous materials.

Some methods, calculation of the accessible surface area and geometric pore size distribution, analysis of the structural connectivity and percolation analysis of the porous space are explained in Ref. [33].

Results for the relations among porosity, density, and dynamics in porous lithium silicate systems can be found in Chapters 6 and 7.

Some of these values have correlated with each other. For example, as already discussed in Chapter 2, porosity is a function of the fractal dimension of holes. Since the porosity is a function of density (see Chapter 6), it means that the fractal dimension of the hole is a function of the density as schematically shown in Fig. 2.12 in Chapter 2. The similar relations should hold among density, packing fraction and fractal dimension of the structure, at least for the monofractal case.

## 4.8 Structural Information Obtained by Molecular Dynamics Simulations—With Examples of NaCl, Silicates, and Ionic Liquids

### 4.8.1 Pair Correlation Functions

Liquids have homogeneous structure at longer length scales and characteristics of them can be well represented by the function based on the radial distance $r$.

Pair correlation function [1, 2] is a fundamental function of the theory of liquids [34, 35] and can also be used for characterizing supercooled liquids and glasses.

From MD simulation, the pair correlation function among different kind of particles $i$ and $j$ is obtained by counting the number of $j$ particles within the shell with width $r$ at distance $r$ from particle $i$, $n_j(r - \Delta r/2, r + \Delta r/2)$.

$$g_{ij}(r) = \frac{V}{N_i \cdot N_j} \left\langle \sum_{j=1}^{N_j} \frac{n_j(r - \Delta r/2,\ r + \Delta r/2)}{4\pi r^2 \Delta r} \right\rangle, \tag{4.8}$$

where $V$ is a volume of the system, $N_i$ and $N_j$ are the numbers of the species $i$ and $j$ and the term $4\pi r^2$ is for the surface area of the sphere.

When the species $i$ and $j$ are the same one, the function is

$$g_{ii}(r) = \frac{V}{N_i \cdot (N_i - 1)} \left\langle \sum_{i=1}^{N_i} \frac{n_i(r - \Delta r/2,\ r + \Delta r/2)}{4\pi r^2 \Delta r} \right\rangle, \tag{4.9}$$

where now $n_i(r - \Delta r/2, r + \Delta r/2)$ is the number of $i$ particles within the shell with width $\Delta r$ at distance $r$. The $(N_i - 1)$ term is the number of surrounding particles, which does not include the central particle $i$.

Practically, the function can be calculated during the MD run and it is useful to check the status of the calculation.

Pair correlation functions can be calculated among ions, atoms, molecules. For example, in the case of ILs with internal structures, the function $g(r)$ for each pair can be determined for the center of mass position (or center of charges) of each ion. These functions become 1 when the mean density is attained.

If one calculates the function during MD runs, distances among $i$ and $j$ particles will randomly appear, and this may prevent the effective coding for parallel computations. A coding with random access may be better to put it out of the loop of the calculation of force, if necessary.

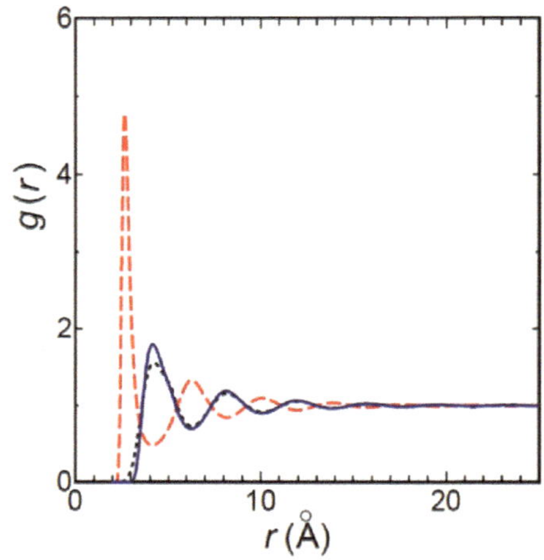


**Figure 4.1** An example of the correlation function, g(r). Data are taken from the result of molecular dynamics simulation for molten NaCl at 1500 K using the same potential parameters as in Chapters 8 and 9. Dashed (red) curve: Na–Cl pair, Dotted (black) curve: Na–Na, Solid (blue) curve: Cl-Cl pair.

The structure thus examined can be used to obtain the more coarse-grained model for the simulations with keeping the characteristics of the structures in the atomistic-level. Of course, it is also possible to calculate $g(r)$ for each atomic species or for partial structures if necessary. Starting from pioneer works by Fumi and Tosi [36, 37], nowadays many kinds of potential parameters are available for alkali halides including polarizable ion models [38]. Researchers can select suitable ones by the required quality, calculation cost, *etc.*, to satisfy the objective of each work. In Fig. 4.1, an example of pair correlation functions for molten NaCl obtained by classical MD simulations is shown.

Here we used the potential parameters in [39], for which NaCl–water interactions in Chapters 8 and 9 are calculated so that results can be compared with colloidal solutions in these chapters. As usual, molten NaCl, Na–Cl pair shows a sharp peak with the shortest distance, while Na–Na and Cl–Cl pairs show quite similar functions to compensate for the charges in Na–Cl pairs.

### 4.8.2 Running Coordination Number and Fractal Dimension Analysis

The running coordination number $N(r)$ is related to the pair correlation function $g(r)$ as follows when $i$ and $j$ are different species [2, 3].

Here, $n$ is chosen such that $n\Delta r$ becomes $r$, and $< >$ indicates an ensemble average.

$$N(r) = < 4\pi r^2 \sum_n \frac{N_j}{V} g_{ij}(n\Delta r) > \tag{4.10}$$

The coordination number of $j$ atoms around $i$ atom can be obtained as a function of the distance $r$ and the first coordination shell is often defined by the value at the position of the first minimum of the pair correlation function of the $i$–$j$ pair.

From the running coordination number, one can determine the fractal dimension, $d_f$, of the structure as follows:

$$d_f = \frac{d \log N(r)}{d \log r} \tag{4.11}$$

If the system consists of several domains or hierarchy structures, more than one fractal dimension will be obtained as shown for several cases of gels in Chapter 3. The coexistence of several exponents can also be analyzed by the multifractal analyses as shown later.

The fractal dimension of the structure affects the motions on this structure, and it can be determined from the trajectories of any units of structures such as atoms, ions, and molecules.

### 4.8.3 Charge Radial Distribution Function

An alternative arrangement of cation and anion in the ionic system tends to neutralize structure at a certain length scale. This behavior is regarded as screening in the strongly coupled system such as NaCl [40, 41] and charge radial distribution in the following form is suggested.

$$q(r) = \frac{A}{r}\exp(-r/\lambda_D)\sin(2\pi r/d + \phi), \tag{4.12}$$

where $A/r$ is the amplitude, $\varphi$ is the phase shift, $d$ is the period of the oscillations the $\lambda$ corresponds to the Debye length.

In ionic structures, and one can consider the charge distribution function $Q(r)$ [41–43] by assuming that the charge is simply on the center of mass (or charge) position,

$$Q(r) = Q_+(r) + Q_-(r) = \{g_{++}(r) + g_{--}(r) - 2g_{+-}(r)\}e \tag{4.13}$$

For monovalent systems such as NaCl with formal charges, $e$ (a weight by a unit charge) is 1. The function is related to layers of charges (charge density wave, CDW). In this function, the mean value for the structure around cation and that around anion was taken. This is reasonable for example in molten NaCl, for which $g(r)$ for Na–Na pair is comparable to that for the Cl–Cl pair. However, if such structures are modified by other components, sometimes each central ionic species need to be distinguished. This function decays to 0 when the charge neutralization is attained at a certain distance. Function concerned with the density wave (DW) regardless of charge distribution can be considered even for the ionic system. The density distribution function $G(r)$ is defined by the following equations.

$$G(r) = G_+(r) + G_-(r) = (g_{++}(r) + g_{--}(r) + 2g_{+-}(r)/4 \tag{4.14}$$

Typical length scales for these functions are informative to learn the difference of the main factors controlling them. This function becomes 1 when the value becomes the mean density. If we compared the $Q(r)$ and $G(r)$ of the molten NaCl, as in Fig. 4.2, one can see the CDW continues longer than that for the

DW. The first peak in CDW is affected by the DW and not necessarily determined by charge ordering alone.

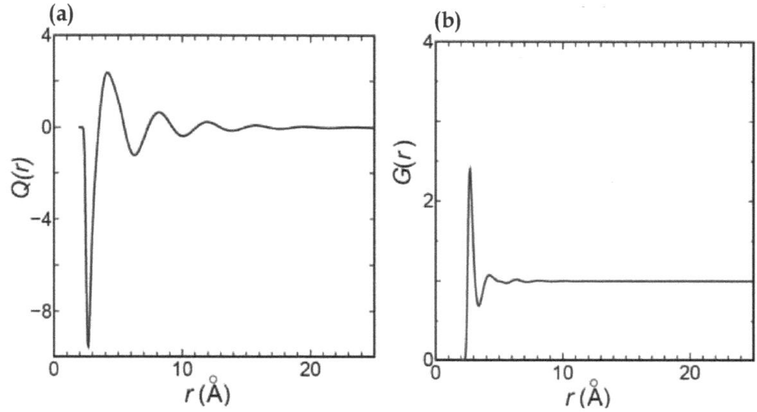


**Figure 4.2** (a) Examples of charge distribution function, $Q(r)$ and (b) density distribution function, $G(r)$. The same run for molten NaCl shown in Fig. 2.1 was used. The charge density wave shows a longer length scale than that in the density wave.

### 4.8.4 Determination of Debye Length from Charge Density Wave (CDW)

If the plots of $\ln|Q(r)r|$ against $r$ have a straight line when the maxima of the peaks (envelopes) are connected, the characteristic length $\lambda_Q$, which corresponds to the Debye length for the screening of the Coulombic term in the simple dilute ionic systems, can be determined from the slope, $-1/\lambda_Q$. Such analyses have been used to examine the length scales of ILs near the glass transition temperature [43] to show the contribution of different length scales. The method was applied to learn the length scale associated with the structure of salts in water (and water in salt) with different concentrations [44]. Figure 4.3 is an example of the function in the molten NaCl, where Coulombic interaction was calculated by the Ewald method. In this case, oscillation with the comparable widths and intervals continues for a long length scale. Caution should be paid for the physical meaning of the Debye length, since the values in dense ionic systems are not necessarily the same one as in a dilute system. The interaction observed is a renormalized one by

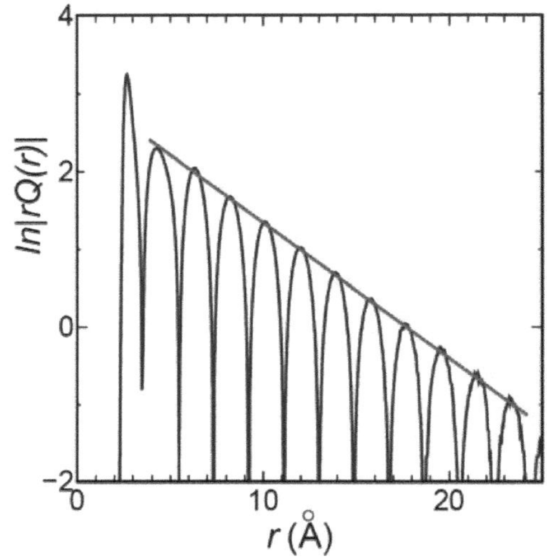


**Figure 4.3** An example of the plot of ln|$rQ(r)$| against $r$. Data are for molten NaCl at 1500 K. From the slope connecting the maxima of peaks (envelopes), one can determine a Debye length, although the physical meaning of the dense ionic systems is not necessarily the same as in that in dilute solution.

the interactions from other ions. The Debye length determined from the slope is found to be 5.94 Å and the value in the form of Eq. 4.3 means the length at which the $|rQ(r)|$ value is reduced to $1/e$ (~0.36788) times of the initial value. Of course, it does not mean the correlation is lost completely at this length. Correlation in Fig. 4.3 continued to cover whole observed regions ~24 Å or longer. Since the function is rapidly decaying, good statistics are required to discuss this kind of length scales.

### 4.8.5 Screening Length in Interfacial Systems

Screening at ionic surfaces is an interesting problem concerning with this structure of the CDW. Desai [40] has examined the system of NaCl confined between two slabs (neutral and charged ones). The function $Q(z) = \rho(z)^+ + \rho(z)^-$, where $z$ is a distance measured from the surface and following one-dimensional form is successfully applied for both octopolar (111) and dipolar (110) surfaces.

$$Q(z) = A\exp(-z/\lambda_Q)\sin(2\pi z/d + \phi), \tag{4.15}$$

where $A$ is amplitude, $\lambda_Q$ is the characteristic length corresponds to the Debye length. Here $d$ and $\phi$ are the period of oscillation and phase shift, respectively.

Large effects of surfaces are reported, and their results suggest that the role of the apparatus used in the experiments cannot be neglected.

## 4.9 Characterization of Complex Systems

The meaning of complex systems discussed is sometimes confusing. The systems having different scales, components are complex. On the other hand, the word "complex system" can be used to mean the existence of deterministic Chaos, which shows complex behaviors even though the system, is represented by quite simple equations. In this book, "complex" is simply concerned with the first meaning, where the multifractal mixing of length scales resulted in heterogeneous structures, although the second meaning is important to consider the origin of behaviors of heterogeneous motions. Here the word "complex" also means the modifications caused by nanostructures.

### 4.9.1 Complexity of the Systems and Vitrification of Systems

From the structural point of view, complexity of systems due to coexistence and/or mixing of different kinds of structural units, length scales, different sizes of domains are often found. One of the causes of the complexity of the structures (and dynamics related to them) of systems is the existence of a plural number of components.

Such complexity prevents the rapid crystallization of the system and that's why usually binary systems are used to study the glass transition, although even in the simple soft sphere model with $r^{-n}$ type potential, the existence of substructures (such as fcc like and bcc like structures) makes the system complicated one to allow it vitrifying.

### 4.9.2 Structures of Silica and Silicates

As well known in mineralogy, almost of crystalline silica systems consist of $SiO_4$ units. Networks formed by tetrahedral units of

$SiO_4$ are commonly found in silica and silicates. Structures formed by them are quite complex ones. Even in silica, which consists of Si and O only, a variety of structures with different kinds of connections of units by vertices or by faces are found.

To characterize silica and the silicate systems, the concept of "fragility" is proposed by Angell to characterize the glass-forming liquids [45]. The fragile system shows super-Arrhenius behavior of the dynamics, while the strong system shows temperature dependence closer to the Arrhenius behavior.

Simple silicates are binary mixtures of silica and oxides such as $M_2O$ (M=Li, Na, K—) and fragility of the system is known to be affected by the mixing of another component. Usually, silicate with higher contents of $M_2O$ is known to be more fragile. Therefore, first, we will consider the changes caused by the additional component in structure and dynamics.

The addition of $M_2O$ cut the structures of silica part and it resulted in the decrease of the dimension of networks of $SiO_4$ units with the formation of different substructures as shown by different $Q_n$ structures (see Section 4.11) as well as different statistics of rings. The distribution of coordination numbers in $MO_x$ units also brings some complexity.

Specific conditions used in MD simulations of porous lithium silicate systems including the original glassy systems can be found in Chapters 6 and 7 and their appendixes. Similar conditions can be used in many cases.

## 4.10 Existence of Different Length Scales in Ionic Systems

From the dynamical point of view, the coexistence of slow and fast particles is related to that of different length scale motions. Their different temperature dependences resulted in the curvature of the diffusivity against $1/T$ plot [4]. That is, super-Arrhenius behavior is caused by the coexistence of different length scales. This view is previously obtained from the changes in the dynamics observed in ionic liquid, EMIM-$NO_3$ [46]. In that paper, it was concluded that the large "fragility," shown by the curvature of diffusion coefficients, can be explained by the changes in the slopes in both temperature regions rather than the suddenly caused rapid decrease of dynamics near $T_0$. Above the inflection point, long-ranged motions with cooperative motions are dominant, while below it, the localized motions become dominant. Then, the acceleration of the dynamics with heterogeneity near the inflection points, which resulted in the gentle slope in the high temperature region, ensures the large changes in the slopes. As found in the above discussions, the fragility of glass-forming liquids is closely related to the concept of heterogeneity of dynamics. Accordingly, it is suggested that multifractality [46–48] due to the coexistence of the different length scales is a good measure of the fragility as well as the heterogeneity. The concept of multifractal can be seen in the Cantor set shown in Fig. 4.4. As depicted in this figure, the mixing of different exponents resulted in the formation of complex heterogeneous structures.

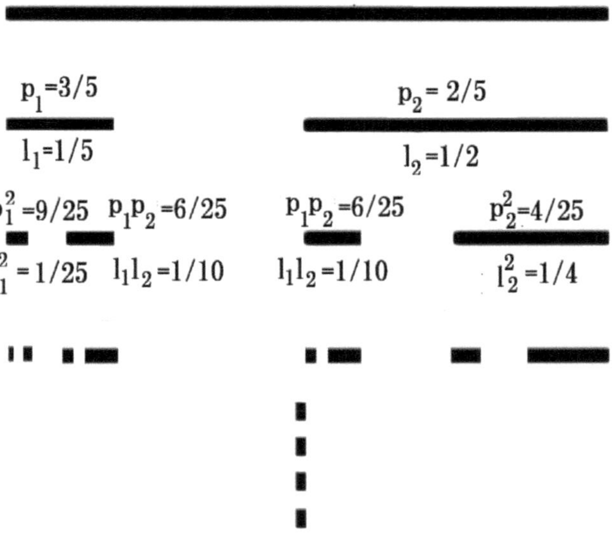


**Figure 4.4** An example of the multifractal Cantor set. When different exponents are mixed as in this figure, the final structures are heterogeneous ones. The length 1/5 and 1/2 remained in the first stage, where the measure is 3/5 and 2/5, respectively. This procedure was repeated. Resultant distribution after $n$ stages is heterogeneous. That is, multifractal analysis gives a spectrum characterizing the heterogeneity. See [3, 21–23] for more detail. Reprinted with permission from Habasaki, J., León, C., Ngai, K. L. (2017), *Dynamics of Glassy, Crystalline and Liquid Ionic Conductors: Experiments, Theories, Simulations*, Springer International.

In the following sections, multifractal analyses have been applied to characterize the paths or sites in ionic systems using density maps [43, 46–48]. It is noteworthy that such maps are obtained by the cumulative positions of particles during MD runs so that it reflects both the mean structures and dynamics governing them.

### 4.10.1 Multifractality of Jump Path and Sites-Singularity Spectra, $f(\alpha)$

Structures of systems formed such different length scales are complicated ones with multifractal nature. In lithium metasilicate, singularity spectra ($f(\alpha)$) [27, 28] has been calculated by the direct method [27] from cumulative positions of $Li^+$ ions during MD run as shown below.

In Fig. 4.5a, the contour map for the cumulative density of Li ions in lithium metasilicate glass at 700 K is shown as an example. Several kinds of substructures such as local ion sites and a pair of ion sites, jump paths connecting several ion sites and each sub structure, have different fractal dimensions. Such a complicated pattern was formed by the coexistence of fast and slow particles (ions). The following characteristics can be seen from the figure. As shown by the color code bar, gradual changes of the color mean those in the density. That is, red color means localized ion sites, while dark blue means the region separated by frameworks made of $SiO_4$ units. Naturally, the structure formed by localized motion has a high density, while the delocalized motion has a lower density. This heterogeneous character is also observed in depths of ion sites. As found in this figure, structures formed by ions are heterogeneous one with many substructures, such as ion sites and ion channels.

It may be worth to mention that this map is obtained for the run during 1 ns and gradual changes of the pattern was found in the following 1 ns and therefore, these paths and sites are not fixed ones at least in the lithium metasilicate at 700 K. The system with high alkali metal ions has a small number of free oxygen atoms, which can modify the network structures gradually, while the network structures in low alkali content glasses are more stable.

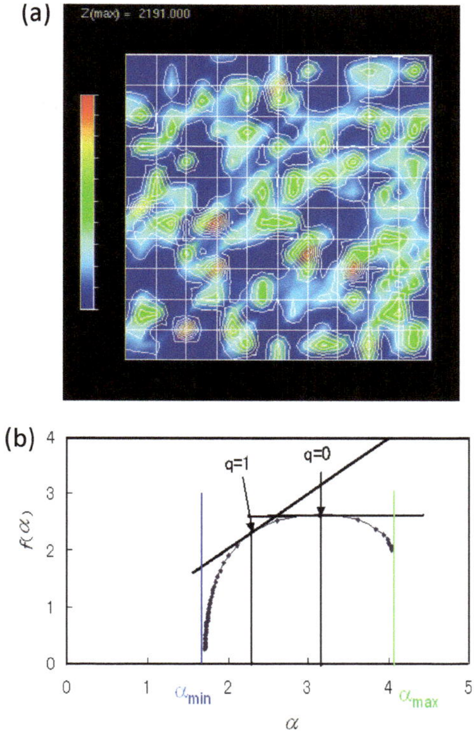


**Figure 4.5** (a) Example of a contour map of accumulated density for Li ions in lithium metasilicate at 700 K. Dense part is due to localization of ions. Such ion sites and jump paths have different fractal dimensions. (b) Example of the corresponding singularity ($f(\alpha)$) spectrum obtained from a 3-D structure of accumulated density. A convex curve means the multifractality. The $\alpha_{\max}$ and $\alpha_{\min}$ correspond to the most rarefied part (delocalization of ions) and the densest part of the density map (localization of ions), respectively, so that the multifractal spectra can represent the heterogeneity of the structures and dynamics. The value at $q = 0$ corresponds to the capacity dimension, while the value at $q = 1$ corresponds to the information dimension. Structures with different fractal dimensions as shown in (a) can be represented by a single curve. Reprinted with permission from Habasaki, J., León, C., Ngai, K. L. (2017), *Dynamics of Glassy, Crystalline and Liquid Ionic Conductors: Experiments, Theories, Simulations*, Springer International.

For characterizing the spatial heterogeneity of structures, a multifractal analysis by using a singularity ($f(\alpha)$) spectrum is a useful method, where many fractal dimensions of substructures can be represented as a single curve. By using the spectra, localization and delocalization shown in the density map can be quantitatively represented.

In Fig. 4.5b, the singularity spectrum obtained by the multifractal analysis for this density map is shown. The convex shape of the spectrum means the multifractality. Generally, the value at $\alpha_{\max}$ corresponds to the rarefied part of the density map and hence meaning delocalization, while $\alpha_{\min}$ corresponds to the densest part of the density map meaning the localization. Therefore, the difference of these values can be used as a measure of the strength of the heterogeneity. The value at $q = 1$ corresponds to the capacity dimension and when the value is larger than the percolation threshold value for 3-D (=2.53 [49]), it means that the jump path is percolated. One should note that the value larger than the percolation threshold is the essential requirement for the occurrence of the diffusion or conduction within an observation time.

### 4.10.2 Calculation of the Singularity Spectrum

Here the method of how to calculate the singularity spectrum [27] is summarized. Singularity spectrum, $f(\alpha)$, is obtained from the probability $\mu_i(\delta)$.

The procedure to obtain the function from the (accumulated) density profile of Li ions is described as follows.

A pattern (a density map) was divided into cells with size $\delta$.

Probability $\mu_i(\delta)$ (=$n_i/\Sigma n_i$) of the cell is measured. In this case, the number "how many times ions visit the $i$-th cell during the run" is regarded as $n_i$.

The measure $\mu_i(\delta)$ is represented as

$$\mu_i(\delta) = \delta^{\alpha}, \tag{4.16}$$

where $\alpha$ is the strength of the singularity. The probability of $\alpha$ lying between $\alpha'$ and $\alpha' + d\alpha'$ is

$$\rho(\alpha') \propto \delta^{-f(\alpha')} d\alpha'. \tag{4.17}$$

Thus, the multifractal spectrum, $f(\alpha)$ corresponds to the fractal dimension of the set of small cells with a singularity strength $\alpha$.

Each spectrum is obtained by using the $q$-th moment.

The concentrated region is emphasized in $[\mu_i(\delta)]^q$, when $q$ is positive, while the rarefied region is emphasized when $q$ is negative.

When $q = 0$, $f(\alpha(q))$ value corresponds to the capacity dimension, $D_0$, while when $q = 1$, the value corresponds to the information dimension $D_1$. Correlation dimension is given by $D_2$ with $q = 2$.

Using the normalized $q$-th moment, $\mu_i(q,\delta)$, the numerical values of $\alpha(q)$ and $f(\alpha(q))$ are obtained for a certain range of $q$ ($-40 < q \leq 30$ were used in our analysis) for several $\delta$ values using the following relations:

$$\alpha(q) = \sum_{i=1}^{N} \mu_i(q,\delta) \ln \mu_i(\delta)/\ln(\delta/L), \tag{4.18}$$

$$f(\alpha(q)) = \sum_{i=1}^{N} \mu_i(q,\delta) \ln \mu_i(q,\delta)/\ln(\delta/L), \tag{4.19}$$

where $\mu_i(q,\delta)$ is defined by

$$\mu_i(q,\delta) = [\mu_i(\delta)]^q / \sum_{i=1}^{N} [\mu_j(\delta)]^q. \tag{4.20}$$

Care must be taken for the numerical errors in calculations because wide ranges of $q$ values are treated.

The obtained spectra are useful for understanding the dynamic heterogeneity forming the density profile. A convex shape of the curve in the $f(\alpha)$ spectrum means the multifractality (mixing of more than one exponent). That is, the structure is formed by the mixing of localized and accelerated motions having different exponents.

Using the singularity spectra, one can compare the heterogeneity of different systems on the same basis. This is because these relations have a similar structure as the thermodynamic formalism of equilibrium thermodynamics, which can be a general framework to study the dynamic and structural heterogeneity.

In Fig. 4.6, a density map for $EMIM^+$ ion in the ionic liquid, 1-ethyl-3-methylimidazolium nitrate ($EMIM\text{-}NO_3$) obtained by MD simulations and corresponding $f(\alpha)$ spectrum are shown.

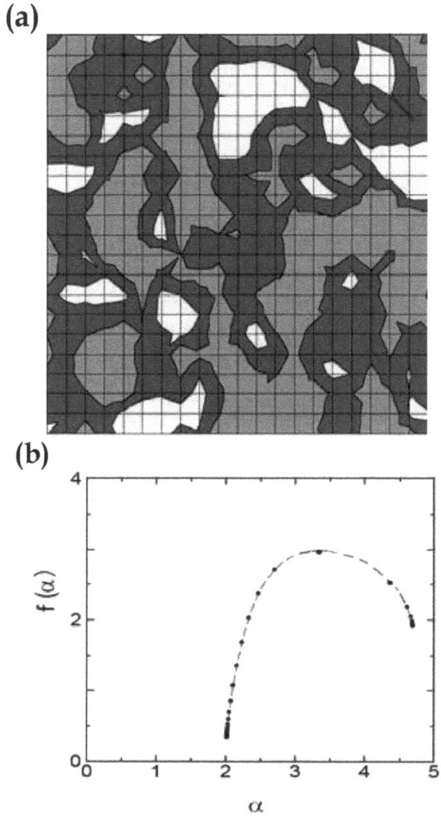


**Figure 4.6** (a) An example of a density map of $EMIM^+$ ions of $EMIM-NO_3$ at 370 K projected on a plane. Colors are changed by a logarithmic scale. (b) An example of the singularity spectrum for the cumulative density of $EMIM^+$ ions in the same system obtained from a 3-D structure of accumulated density. Convex shape of the curve means the multifractality. The maximum position with $q = 0$ corresponds to the capacity dimension. The value 3 means the three-dimensional connections of the paths in the liquid state. The $\alpha_{max}$ and $\alpha_{min}$ correspond to the most rarefied part and the densest part of the density profile, respectively. Reprinted with permission from Habasaki, J., León, C., Ngai, K. L. (2017), *Dynamics of Glassy, Crystalline and Liquid Ionic Conductors: Experiments, Theories, Simulations*, Springer International.

In ILs, ions are charge carriers and at the same time glass formers. Therefore, ion sites for cations are separated by moving anions. The sites seem to be connected and show complicated fractal shapes. Even though it has a quite different appearance from that in ionically conducting glasses, quite a similar spectrum in the shape is obtained. While both $\alpha_{min}$ and $\alpha_{max}$ values in the spectrum in IL are larger than the lithium silicate glasses and the capacity dimension, $D_0$, is also larger than those of lithium silicate glasses. These differences are reasonable to consider the IL at 370 K is in the liquid state.

This situation with multifractal is not limited to the ionic systems. Generally, multifractal analysis can treat complex systems represented by different shades. For example, it was used to characterize fractures [50, 51], turbulence [52], reactions over a rough surface [53, 54], genomes [55], and crystalline aggregates formed in a drying gel medium [56].

A porous system is expected to take a role as a storage of gases as well as liquids, and the distributions of oil and gases in porous systems is known to be heterogeneous. Recently, multifractal analyses were used to characterize the pore structure of tight sandstone [57].

Strong correlations were found to exist between the porosity and the multifractal parameters, such as $\alpha_{min}$, $\alpha_{max}$, $\Delta\alpha_R$ $(=\alpha_0-\alpha_{min})$ $\Delta\alpha_L(=\alpha_{max}-\alpha_0)$, AI $(=(\Delta\alpha_L - \Delta\alpha_R)/\Delta\alpha)$, and $\Delta f$ $(=f_{\alpha max} - f_{\alpha min})$. Here $\alpha_0$ is a position of the maximum.

Peng and coworkers proposed a classification of types of pore structures based on these parameters.

Thus, the multifractal analysis seems to be a promising method for systematic comparison of complex systems, the classification of the materials, and characterization of surfaces as well as their porosity.

It will be interesting to use the classificaion of multifractal materials suggested in Chapter 2 for these systems.

The method is also applied to the structures obtained by simulations of non-ionic systems. As demonstrated by Sakikawa and Narikiyo [58], the method was successfully applied to examine the heterogeneity in two-dimensional supercooled liquids of the soft sphere model and found that the cooperative rearrangement regions (CRR) [59, 60] can be identified as the weakly bonded regions of the multifractal structure of (fictive) bonds.

Besides the $f(\alpha)$ spectra, the generalized dimension, $D_q$, can be used for the analyses of multifractality, because the dimension is related to $f(\alpha(q))$ by a Legendre transformation, $\tau(q) = (q-1)D_q$, where $\tau(q)$ and $q$ are conjugate thermodynamic variables to $f(\alpha)$ and $\alpha$.

### 4.10.3 Relationship between "Fractal Dimension of Random Walks," and Power Law Exponent of MSD

In 1967, the concept of fractal dimension of the random walks, $d_w$, was introduced by Mandelbrot to measure the complexity of the coast [20].

One can imagine the situation that one measures the length of the coast on foot with a fixed stride, so that the value for an uneven coastline will be longer compared with the value measured on the small map by a divider. This is because, details of the curves are coarse-grained in the latter. If the coast is more complex, the difference of these values will be larger. Therefore, if one measured a total length and/or total frequency to use the ruler to cover the coast by using several fixed lengths of rulers, one can determine how the coastline is complex.

Of course, coast or trajectory is not an exact fractal in the sense of the self-similarity but has the characteristic of statistical fractal and trajectories obtained by MD directly reflect the motion of particles (atoms, ions, and molecules) and its complexity is a good measure of the dynamics.

Some examples of trajectories of ions found in molecular dynamics simulations of the ionic system are shown in Fig. 4.7.

As shown in Fig. 4.8, using divider (i.e., a straight ruler) of length $L$, one can count how many times, $N$, are required to cover the trajectory.

From the slope in the double logarithm plots, one can determine the fractal dimension of random walk, $d_w$, defined by

$$N = AL^{-d_w} \tag{4.21}$$

Normal random walk like in Brownian diffusion has $d_w = 2$. The value increases if the trajectories are complex and localized, while the value decreases if the trajectories are like straight lines and delocalized. Curves on the trajectory can overlap each other and therefore the obtained fractal dimension is a latent one, which can be larger than the dimensionality of space and this situation is different from the coast.

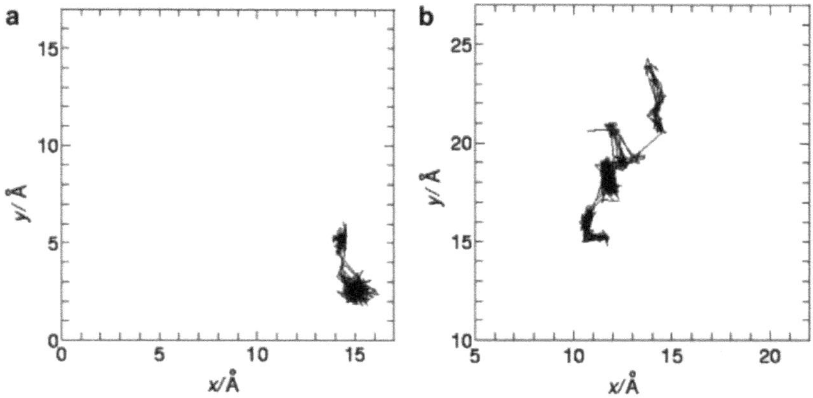


**Figure 4.7** Example of trajectory of a particle obtained by MD for lithium ion in lithium metasilicate at 1000 K, (a) a localized ion and (b) an ion accelerated by forward correlated motions, where the motion is highly cooperative.

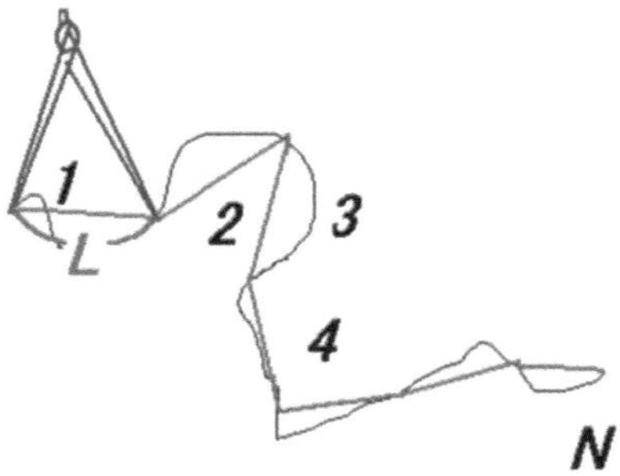


**Figure 4.8** Procedure how to determine the fractal dimension of random walks from the trajectory. Using the divider of length $L$, counts required to cover the trajectory are determined. This procedure has been repeated by changing length $L$. The value $N(L)$ for the sum of all trajectories is plotted against $L$ in a double. From the slope, one can determine the fractal dimension of the random walks. If more than one region is found in the plot, the motion is multifractal rather than monofractal.

Previously, we have reported the existence of two different length scales in trajectories of Li ions in lithium silicate systems [43] because more than one region was found in the (double logarithmic) plots of $N$ against $L$. Similar behaviors are also found in other ionic systems, that is typical ionic liquid (IL), EMIM-$NO_3$ [46].

Due to the existence of two different length scale regions showing different exponents, such walks are characterized as "multifractal (random) walks" [61, 62]. Two length scales are thus represented by two different values of $d_w$. In general, the dynamics of ions in disordered materials show large dynamical heterogeneity both temporally (waiting-time distribution of jumps) and spatially (geometrical correlation among successive motions). The fractal dimension analysis of the trajectories is concerned with the spatial term.

Here the residence time (temporal term) in each site does not affect the result. That is, $d_w$ is determined solely by the geometrical correlations. The slope with $\theta$ (=$2/d_w$) is expected [25] for the fractional power-law region of the MSD (before the onset of the diffusive regime) in the theory of simple fractal. In this case, a temporal term might be included in $d_w$.

If the ionic site is visited by the walk with $d_w$ in a mono-fractal manner, the mean squared displacement in the power-law regime is given by

$$\langle r^2(t)\rangle \sim t^{2/dw}. \tag{4.22}$$

Naturally, the "power law" dependence of the MSD explains the large non-linear effect of the slowing down with a change of these values.

At first sight, an explanation for the mono-fractal case seems to hold in many systems; however, we have observed two different length scales in $d_w$, in lithium metasilicate or IL, EMIM-$NO_3$, and hence the observed exponent $\theta$ is for mean behavior of the $d_{w1}$ and $d_{w2}$. The former is for the short-length-scale motion while the latter is for the longer-length-scale one. The coexistence of two length scales in trajectories means that the walk of ions has a multifractal nature. It seems to be a natural result obtained from the multifractal nature of the density profile. That is, if dynamics are localized, density in localized regions becomes large and if dynamics are delocalized, low-density regions are formed. In Fig. 4.9, an example of the motion of ion (EMIM$^+$ ion in EMIM-$NO_3$) at 370 K is shown with the corresponding singularity spectrum, $D(h)$. The spectrum for the trajectories can be obtained similarly to the $f(\alpha)$

spectrum of the density profile. Multifractality of the walks can be characterized by the singularity spectrum using Hurst exponent [63, 64], $h$, defined by the following equation in the case of one-dimensional motion:

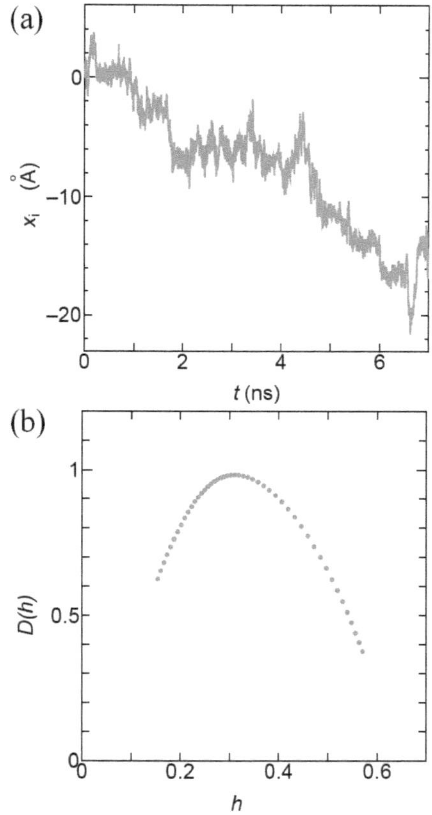


**Figure 4.9** (a) An example of the changes in the position of an EMIM$^+$ ion along X axis at 370 K. The motion consists of fast and slow motions. (b) A singularity spectrum $D(h)$ for the motion shown in (a). A convex shape of the spectrum means the multifractality of the dynamics. Reprinted with permission from Habasaki, J., León, C., Ngai, K. L. (2017), *Dynamics of Glassy, Crystalline and Liquid Ionic Conductors: Experiments, Theories, Simulations*, Springer International.

$$\delta x_i(t_0,\delta t) = |x_i(t_0+\delta t) - x_i(t_0)| \approx (\delta t)^{h(t_0)} \tag{4.23}$$

In the power-law region of the MSD, $\langle r^2(t)\rangle \sim t^{\theta}$, with $\theta = 2H$.

Here $H$ is the Hurst exponent for the mean behavior of particles. Since the fractal dimension of the random walk $d_w$ is related to $\theta$ by $\theta = 2/d_w$, the value of $d_w$ for the mean behavior corresponds to the inverse of the Hurst exponent. That is,

$$H = 1/d_w. \tag{4.24}$$

## 4.11 Examination of Network Structures

Complex systems often have complicated network systems.

Here some of the possible methods, which can be used to characterize networks using MD simulations, are shown. The methods include the $Q_n$ distribution and cluster analyses using the dendrogram. These methods are used to characterize the structure of aggregates and gels in Chapters 8 and 9. Multifractal analysis can be applied to this kind of structure. Of course, pair correlation functions and related functions are also useful as well as angular correlations. Formation of higher-order structures such as rings, sheets is also possible to examine.

### 4.11.1 Distribution of $Q_n$ (or $Q^n$) Structure

To characterize network formed by structural units such as $SiO_4$ in silica or silicates, the distribution of $Q_n$ structures, where $n$ means the number of the bridging oxygen atoms, is often used. Both the subscript and the superscript are found in the literature; however, the subscript $n$ will be used hereafter, because it is convenient to distinguish it from other exponents, which appears many times in this book. An example of such structures found in MD simulations of silicate is shown in Fig. 8.4 in Chapter 8.

The distribution represents not only local structures of $SiO_4$ units but also characteristics of network structures, although further information will be required for how they are packed in the system. The different distribution of $Q_n$ values means that there are different kinds of networks. Distributions are found to be closely related to polyamorphism as found in silicate [31]. Although the distribution is sometimes used to judge the validity of the MD simulation, one should be careful that the distribution depends on the pressure of the system considerably, because each $Q_n$ structure has specific partial volume [30, 31]. Such structures examined for nanosilica colloidal solutions as well as for gels will be examined in Chapters 8 and 9. It is interesting to note that corresponding structures in water are also closely related to the polyamorphism as well as an anomaly of density [65].

### 4.11.2 Fractal Dimension Analyses for 2-D and 3-D Structures

The method used for the calculation of fractal dimension of the random walks is one of the box-counting methods. The method is also applicable to the analysis of other structures. For two-dimensional images obtained by experimental methods such as a transmission electron microscope (TEM), the analyses are often applied. Then the method is concerned with 2-D structures.

One of the possible choices is to analyze 3-D structures obtained for the radial direction. Although the dimension was reduced to radial direction only, results keep the information of 3-D structures and we called it 3-D method when it is compared with the 2-D method based on the 2-D images. It can be done by using pair correlation function, $g(r)$, or the running coordination number, $N(r)$ (see Section 4.8.2). The comparison of these methods for clusters, aggregates, and gel will be shown in Chapter 9. The three-dimensional method successfully characterizes these systems; however, 2-D methods are found not necessarily to be useful for this purpose. On the other hand, both methods can show the existence of different length scale regions.

### 4.11.3 Cluster Analysis by Dendrogram

To represent a hierarchical clustering of structural units, the dendrogram (the tree structure) can be used. A simple example to show the connections among clusters 1–3 are shown in Fig. 4.10. The height at which two clusters are connected represents the distance or dissimilarity between clusters. If we use this analysis for the connection of silica in aggregates, and gels as in Chapter 9, one can understand how the structures are formed from units. In the example of Fig. 4.10, it may suggest that the cluster 1 and 2 are connected first; then the combined cluster (1,2) is connected to 3. However, further confirmation is necessary to say so, because it does not necessarily clarify the ordinal relation.

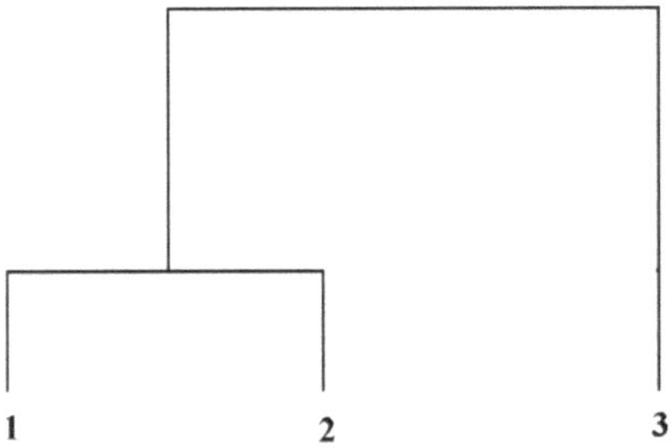


**Figure 4.10** Simple example of dendrogram to show how cluster 1, 2 and 3 are combined to form a larger cluster.

If a different measure of distance is used, a quite different impression might be obtained from the same structure and therefore careful discussion for the result is required.

## 4.12 Conclusion

Some fundamental functions and methods useful to study the simple and complex systems by using MD simulations are summarized in this chapter.

The characteristics of the pair correlation function as well as charge distribution functions representing charge orderings are shown with the arguments of the characteristic length scale related to the Debye length. For arguments of nanostructured systems, fractal dimension analyses are useful to characterize structures. Especially, multifractal analyses will be useful when there is more than one characteristic length in the system.

There are many nanostructured systems with networks Methods useful to characterize the network structures are introduced in this chapter. Distribution of $Q_n$ structures concerning $SiO_4$ units is often used to characterize networks in silica and silicates. The dendrogram (tree structures) of connections of units is useful to understand how such units are connected. Some of these methods will be applied to the analyses of clusters, aggregates, and gel in Chapters 8 and 9.

## References


1. Allen, M. P., Tildesley, D. J. (1989), *Computer Simulation of Liquids*, Oxford University Press, New York.
2. Ueda, A. (2003), Molecular Simulations-from classical to quantum method-, SHOKABO, Tokyo, in Japanese.
3. Habasaki, J., León, C., Ngai, K. L. (2017), *Dynamics of Glassy, Crystalline and Liquid Ionic Conductors, -Experiments, Theories, Simulations*, Springer International.
4. Car, R., Parrinello, M. (1985), Unified approach for molecular dynamics and density functional theory, *Phys. Rev. Lett.*, 55, pp. 2471–2474.
5. Car, R. and Parrinello, M. (1989), "The Unified Approach for Molecular Dynamics and Density Functional Theory", in Simple Molecular Systems at Very High Density, vol. 186 of NATO ASI Series, series B, Physics, Loubeyre, P. P. and Boccara, N. (Eds.) (Plenum Press, NY) pp. 455–476.
6. Habasaki, J., Ngai, K. L. (2015), Rigidity and soft percolation in the glass transition of an atomistic model of ionic liquid, 1-ethyl-3-methyl imidazolium nitrate, from molecular dynamics simulations—Existence of infinite overlapping networks in a fragile ionic liquid, *J. Chem. Phys.*, 142, 164501 (2015).
7. Verlet, L. (1967), Computer "experiments" on classical fluids. I. Thermodynamical properties of Lennard-Jones molecules, *Phys. Rev.* 159, pp. 98–103. doi:10.1103/PhysRev.159.98.
8. Gear, C. W. (1971), *Numerical Initial Values Problems in Ordinary Differential Equations* (Prentice-Hall, Englewood Cliffs, NJ).
9. Ewald, P. P. (1921), Die Berechnung optischer und elektrostatischer Gitterpotentiale, *Ann. Phys.*, 64, pp. 253–287.
10. Darden, T., York, D., Pedersen, L. (1993), Particle mesh Ewald: An N·log(N) method for Ewald sums in large systems, *J. Chem. Phys.*, 98, pp. 10089–10092.
11. Habasaki, J., Ngai, K. L. (2015), Molecular dynamics study of heterogeneous dynamics in lithium disilicate crystal, *J. Electroceram.*, 34, pp. 43–56.
12. Shimada, T., Tomoda, S., Kitamura, T. (2009), *Ab initio* study of ferroelectricity in edged $PbTiO_3$ nanowires under axial tension, *Phys. Rev. B*, 79, 024102 (1–7).
13. Martin, E., Pier, Palla, P. L., Cleri, F., Bouzid, A., Ori, G., Le Roux, S., Boero, M., Massobrio, C. (2018) On the occurrence of size effects

in the calculation of thermal conductivity by first-principles molecular dynamics: The case of glassy $GeTe_4$, *J. Non-Cryst. Solids*, 498, pp. 190–193.

14. Adam, G., Gibbs, J. H. (1965), On the temperature dependence of cooperative relaxation properties in glass-forming liquids, *J. Chem. Phys.* 43, pp. 139–146.
15. Yoshidome, T., Yoshimori, A., Odagaki, T. (2007), Free energy landscape and cooperatively rearranging region in a hard sphere glass, *Phys. Rev. E*, 76, 021506 (1–6).
16. Horbach, J., Kob, W., Binder, K., Angell, C. (1996), Finite size effects in simulations of glass dynamics, *Phys. Rev. E*, 54, pp. R5897–R5900.
17. Habasaki, J., Ueda, A. (2016), Molecular dynamics study of one component soft-core system-analytical expression of non-equilibrium relaxation in constant pressure conditions, *J. Non-Cryst. Solids*, 447, pp. 212–222.
18. Schweigert, I. V., Lehtinen, K. E. J., Carrier, M. J., Zachariah, M. R. (2002), Structure and properties of silica nanoclusters at high temperatures, *Phys. Rev. B*, 65, 235410 (1–8).
19. Habasaki, J., Ishikawa, M. (2014), Molecular dynamics study of coagulation in silica-nanocolloid–water–NaCl systems based on the atomistic model, *Phys. Chem. Chem. Phys.*, 16, pp. 24000–24017.
20. Thorne, K. S. (1980), Multipole expansions of gravitational radiation, *Rev. Mod. Phys.* 52, pp. 299–340.
21. Wolf, D. (1992), Reconstruction of NaCl surfaces from dipolar solution to the madelung problem, *Phys. Rev. Lett.*, 68, pp. 3315–3318.
22. Wolf, D., Keblinski, P., Phillpot, S. R., Eggebrecht, J. (1999), Exact method for the simulation of Coulombic systems by spherically truncated, pairwise $r^{-1}$ summation, *J. Chem. Phys.*, 110, pp. 8254–8282; doi: 10.1063/1.478738.
23. Avendaño, C., Gil-Villegas, A. 2006, Monte Carlo simulations of primitive models for ionic systems using the Wolf method, *J. Mol. Phys.*, pp. 1475–1486.
24. Mandelbrot, B. (1967), How long is the coast of Britain? Statistical self-similarity and fractional dimension, *Science*, 156, pp. 636–638.
25. Ben-Avraham, D., Havlin, S. (2000), *Diffusion and Reactions in Fractals and Disordered Systems*, Cambridge University Press, Cambridge.
26. Mandelbrot, B. (1999), *Multifractals and 1/f Noise: Wild Self-Affinity in Physics (1963–1976)*, Selecta Volume N. Springer.

27. Chhabra, A., Jensen, R. V. (1989), Direct determination of the $f(\alpha)$ singularity spectrum, *Phys. Rev. Lett.,* 62, pp. 1327–1330.

28. Halsey, T. C., Jensen, M. H., Kadanoff, L. P., Procaccia, I., Shraiman, B. (1986), Fractal measure and their singularities: The characterization of strange sets. *Phys. Rev. A,* 33, pp. 1141–1151. doi:10.1103/PhysRevA.33.1141.

29. Greaves, G. N., Sen, S. (2007), Inorganic glasses, glass-forming liquids and amorphizing solids, *Adv. Phys.*, 56, pp. 1–166.

30. Doweidar, H. (1996), The density of alkali silicate glasses in relation to the microstructure, *J. Non-Cryst. Solids,* 194, pp. 155–162. https://doi.org/10.1016/0022-3093(95)00489-0.

31. Habasaki, J., Ngai, K. L. (2013), Molecular dynamics study of network statistics in lithium disilicate: $Q_n$ distribution and the pressure-volume diagram, *J. Chem. Phys.*, 139, 064503 (1–14).

32. For example, see https://reference.wolfram.com/language/ref/Dendrogram.html (The address is confirmed to be valid on 16th march, 2019).

33. Sarkisov, L., Harrison, A. (2011), Computational structure characterization tools in application to ordered and disordered porous materials, *Molec. Sim.*, 37, pp. 1248–1257. DOI: 10.1080/08927022.2011.592832.

34. Chandler, D. (1987), *Introduction to Modern Statistical Mechanics,* Oxford University Press, New York.

35. Hansen, J. P., MacDonald, I. R. (1990), *Theory of Simple Liquids,* 2nd ed., Academic Press, London.

36. Fumi, F. G., Tosi, M. P. (1964), Ionic sizes and born repulsive parameters in the NaCl-type alkali halides—I: The Huggins-Mayer and Pauling forms, *J. Phys. Chem. Solids,* 25, pp. 31–43.

37. Tosi, M. P., Fumi, F. G. (1964), Ionic sizes and born repulsive parameters in the NaCl-type alkali halides—II: The generalized Huggins-Mayer form, *J. Phys. Chem. Sol*ids, 25, pp. 45–52.

38. Ishii, Y., Kasai, S., Salanne, M., Ohtori, N. (2015), Transport coefficients and the Stokes–Einstein relation in molten alkali halides with polarizable ion model, *Mol. Phys.,* 113, pp. 2442–2450.

39. Cygan, R. T., Liang J.-J., Kalinichev, A. G. (2004), Molecular models of hydroxide, oxyhydroxide, and clay phases and the development of a general force field, *J. Phys. Chem. B,* 108, pp. 1255–1266.

40. Desai, T. G. (2007), Molecular dynamics study of screening at ionic surfaces, *J. Chem. Phys.,* 127, pp. 154707–154713. https://doi.org/10.1063/1.2796161.

41. Keblinski, P., Eggebrecht, J., Wolf, D., Phillpot, S. R. (2000), Molecular dynamics study of screening in ionic fluids, *J. Chem. Phys.*, 113, pp. 282–291.

42. Del Pópolo, M. G., Voth, G. A. (2004), On the structure and dynamics of ionic liquids, *J. Phys. Chem. B*, 108, pp. 1744–1752.

43. Habasaki, J., Ngai, K. L. (2005), Multifractal analysis of dynamic potential surface of ion-conducting materials, *J. Chem. Phys.*, 122, p. 214725, https://doi.org/10.1063/1.1924691.

44. Gubarevich, A., Habasaki, J. (2017), Long-range charge order and electrostatic screening in ionic liquids and concentrated electrolyte solutions, *8th International Discussion Meeting on Relaxations in Complex Systems* (IDMRCS8) July 23–28(25) Wisla, Poland.

45. Angell, C. A. (1998), Thermodynamics: Liquid landscape, *Nature,* 393, pp. 521–524.

46. Habasaki, J., Ngai, K. L. (2008), Molecular dynamics study of the dynamics near the glass transition in ionic liquids, *Anal. Sci.*, 24, pp. 1321–1327.

47. Habasaki, J., Ngai, K. L. (2006), Heterogeneity, multi-fractality and cooperativity in the ionically conducting glasses, *Flow Dyn., IAP Conference Proc.,* 832, pp. 199–202.

48. Habasaki, J., Ngai, K. L. (2011), Multifractal nature of heterogeneous dynamics and structures in glass forming ionic liquids, *J. Non-Cryst. Solids.*, 357, pp. 446–453.

49. Stauffer, D., Aharony, A. (1994), *Introduction to Percolation Theory*, CRC Press.

50. Stach, S., Cybo, J., Chmiela, J. (2003), Multifractal description of fracture morphology: Theoretical basis, *Mater. Charact.,* 51, pp. 79–86.

51. Herrmann, H. J., Hansen, A., Roux, S. (1989), Fracture of disordered, elastic lattices in two dimensions, *Phys. Rev. B,* 39, pp. 637–648.

52. Kestener, P., Arneodo, A. (2004), Generalizing the wavelet-based multifractal formalism to random vector fields: Application to three-dimensional turbulence velocity and vorticity data, *Phys. Rev. Lett.*, 93, 044501 (1–4).

53. Yan, C.-C., Chaudhari, A., Lee, S-L. (2002), Dynamic scaling of diffusion limited reactions over fractal surfaces: Computer simulation and theoretical investigation, *Appl. Surf. Sci.,* 196(1–4), pp. 375–382.

54. Chaudhari, A., Yan, C.-C., Lee, S.-L. (2003), Multifractal scaling analysis of autopoisoning reactions over rough surface, *J. Phys. A: Math & Gen.*, 36, pp. 3757–3772.

55. Oiwa, N. N., Glazier, J. A. (2002), The fractal structure of the mitochondrial genomes, *Physica A,* 311, pp. 221–230.

56. Dutta, T., Giri, A., Choudhury, M. D., Tarafdar, S. (2013), Experiment and simulation of multifractal growth of crystalline NaCl aggregates in aqueous gelatin medium, *Colloids Surf. A*, 432, pp. 127–131.

57. Peng, J., Han, H., Xia, Q., Li, B. (2018), Evaluation of the pore structure of tight sandstone reservoirs based on multifractal analysis: A case study from the Kepingtage Formation in the Shuntuoguole uplift, Tarim Basin, NW China, *J. Geophys. Eng.,* 15 pp. 1122–1136.

58. Sakikawa, W., Narikiyo, O. (2003), Multifractal analysis of the dynamical heterogeneity in a two-dimensional supercooled liquid, *J. Phys. Soc. Jpn.*, 72, pp. 450–451.

59. Adams, G., Gibbs, J. H. (1965), On the temperature dependence of cooperative relaxation properties in glass-forming liquids, *J. Chem. Phys.*, 43, pp. 139–146.

60. Odagaki, T., Ekimoto, T. (2007), Fast and slow relaxations in the free energy landscape, *J. Non-Cryst. Solids,* 353, pp. 2928–3931.

61. Dräger, J., Bunde, A. (1996), Multifractal features of random walks and localized vibrational excitations on random fractals: Dependence on the averaging procedure, *Phys. Rev. E*, 54, pp. 4596–4602.

62. Barcy, E., Delour, J., Muzy, J. F. (2001), Multifractal random walk, *Phys. Rev. E,* 64, 026103 (2001).

63. Feder, J. (1988), *Fractals*. New York: Plenum Press.

64. Zelenyĭ, L. M., Milovanov, A. V. (2004), Fractal topology and strange kinetics: From percolation theory to problems in cosmic electrodynamics, *Phys. Uspekhi*, 47, pp. 749–788.

65. Habasaki, J. (2019), Heterogeneous–homogeneous transition and anomaly of density in SPC/E water examined by molecular dynamics simulations, *Physica A*, 527, 121391.

# Chapter 5

# Molecular Dynamics Simulations of Ionic Motions: Dynamic Heterogeneity as a Basis of Studies of Nanostructured Materials

In this chapter, the mechanism of ionic motions in silicate systems derived from MD simulations will be discussed. The time development of mean-squared displacement (MSD) of ions [1, 2] (corresponding to frequency dependence of ionic motion [1–7]) shows several characteristic time regions [8, 9]. Caged ion dynamics in early time (nearly constant loss region (NCL)) region take important roles in the enhancement of dynamics found in porous lithium disilicate in the *NVE* condition, and details of short-time dynamics will be discussed in [8, 9] and Chapters 6 and 7. Behaviors of characteristic times found in the MSD of ions can be a basis to understand how dynamics are affected by further modifications such as introducing pores. That is, they are useful to distinguish possible mechanisms.

*Molecular Dynamics of Nanostructures and Nanoionics: Simulations in Complex Systems*
Junko Habasaki

ISBN 978-981-4800-77-8 (Hardcover), 978-1-003-04490-1 (eBook)
www.jennystanford.com

## 5.1 Characteristics of Ionic Motions in Ionically Conducting Melts and Glasses

Dynamics of ions are the complex ones having many questions to be solved and are continued to be discussed in papers [1–5, 10], reviews [11–14], and books [6, 7]. In recent works, several points have been addressed and made clearer, where molecular dynamics simulations play roles to gain a nanoscopic understand of these complex behaviors.

One of the challenging problems is the relationships between AC conductivity and DC conductivity, where many functional forms are assumed empirically. Another interesting problem in ionics is the mechanism of the "mixed alkali effect" [6, 7, 15–20], which is extensively examined by molecular dynamics simulations [21–25] as well as experiments and theoretical methods.

In this chapter, we focus on the former phenomena observed in single alkali silicates and related systems. As already mentioned, these results are important as a basis to consider the effect of modification of Li ions by introducing pores, as will be discussed in Chapters 6 and 7.

In experiments, it is well known that conductivity spectra show complex behaviors and let's start a discussion from this behavior.

An example of this behavior is shown in Fig. 5.1 [4, 5]. The ac conductivity $\sigma'(\omega)$ of ionic systems shows a constant value at low frequencies. The value is called dc conductivity. In higher frequency, several regions are observed. The higher frequency region than $\nu_{\times 1}$ showing up in the conductivity spectrum and/or susceptibility is called nearly constant loss (NCL) region. Details of the assignment of frequencies characterizing regions will be explained later.

From this kind of common observations, the real part of the conductivity is often expressed as follows [26–28]:

$$\sigma'(\omega) = \sigma_{dc} + A\omega^n, \tag{5.1}$$

where $\sigma_{dc}$ is the dc conductivity, $A$ is a temperature-dependent parameter and $n$ is a fractional exponent which usually lies between 0.6 and 1. In this expression, ac and dc seem to be independent and overlap.

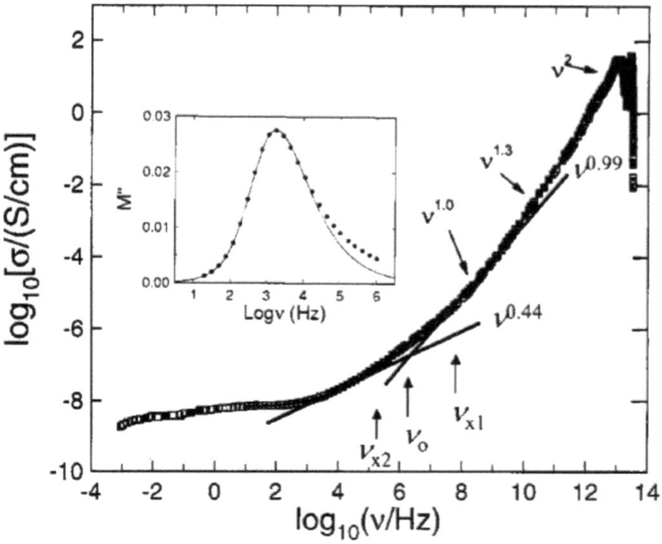


**Figure 5.1** $\text{Log}_{10}\,\sigma'(\nu)$ versus log (frequency/Hz) plot of data at $T$ = 293 K of 0.80LiF-0.20Al$(PO_3)_3$ obtained by Kulkarni et al. (Kulkarni, A. R., Myer, F., Schnider, U., Lunkenheimer, P., Loidl, A. (1999), Ionic Conductivity and Relaxations in Li and (Li and K) Fluorophosphate Glasses, Ceramic Transactions, 92, p. 115) to show the nearly linear frequency dependence of the NCL by the line with slope equal to 0.99. The line with slope 0.44 is the high-frequency dependence of $\log_{10}\,\sigma'(\nu)$ corresponding to the Kohlrausch fit to the electric modulus data (shown in the inset) with $n$ = 0.44. The dashed line in the main figure is the fully cooperative ion conductivity relaxation. The crossover frequencies $\nu_{x1}$ and $\nu_{x2}$ as well as the independent hopping frequency $\nu_0$ are determined from the data in the same manner as explained in Fig. 5.1 (in the following reference). Reprinted from Ngai, K. L., Leon, C. (2002), Cage decay, near constant loss, and crossover to cooperative ion motion in ionic conductors: Insight from experimental data, *Phys. Rev. B*, 66, 064308, pp. 1–11.

Combination of this expression with Kramers–Kronig relations leads to the frequency dependence of the complex conductivity, which can be written as

$$\sigma^*(\omega) = \sigma_{dc}[1 + (j\omega/\omega_c)^n] + j\omega\varepsilon_0\varepsilon_\infty \tag{5.2}$$

This expression is explained as the sum of high-frequency permittivity term, $j\omega\varepsilon_0\varepsilon_\infty$, due to the contributions from all the polarization at higher frequencies, and the power-law frequency dependence (at $\omega > \omega_c$), where $\omega_c$ representing a crossover frequency from the dc conductivity plateau (at $\omega < \omega_c$).

Many experimental data exhibiting such behaviors can be found in books [6, 7]. Origin of such universal behavior of the frequency dependence of the conductivity has been a target of debates for a long time [28–31].

### 5.1.1 Mean-Squared Displacement and Characteristic Time Regions

In molecular dynamics simulations, mean-squared displacement (MSD) of ions or atoms can be obtained using the following relations:

$$<r^2(t)> = \left\langle \left\{ \sum_{i=1}^{N} (\mathbf{r}_i(t) - \mathbf{r}_i(0))^2 \right\} / N \right\rangle, \tag{5.3}$$

where, $\mathbf{r}_i$ is the positional vector of the $i$-th particle.

To learn the origin of these behaviors in the frequency domain, it is useful to examine the time development of mean-squared displacement of ions.

MSD of ions as a function of time shows several distinctly different regions with different functional forms. Figure 5.2 shows the typical behaviors of ionic motion, that is, MSD of Li ions for lithium metasilicate at 700 K obtained by MD simulation. In this figure, characteristic times, $t_{x1}$ and $t_{x2}$, are defined for representing changes in the functional forms in MSD. The value $\tau_0$ obtained based on the following relation of the coupling model [4, 5] is also shown for comparison.

$$\tau = [t_c^{-n} \tau_0]^{1/(1-n)} \tag{5.4}$$

This is a representative example of time (or frequency)-dependent dynamics, where the method can cover the wide time ranges from ~fs to several ~tens ns.

The complex conductivity $\sigma*(\omega)$ is in the frequency domain can be connected to the MSD in the time domain by the following equation [32, 33]:

$$\sigma^*(\omega) = -\omega^2 \frac{N_\rho q_c^{\ 2}}{6H_R kT} \int_0^\infty <r^2(t)> e^{-i\omega t} dt, \tag{5.5}$$

where $N_\rho$ is a number density of mobile ions, $q_c$ (= $Z_i$ $e$) is a charge of the ion, $k$ is the Boltzmann constant, $H_R$ is the Haven ratio [34], and $T$ is temperature.

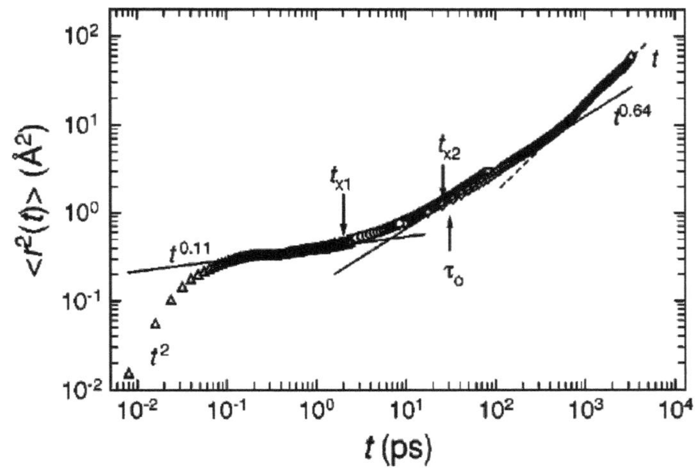


**Figure 5.2** Mean square displacement of Li ions in lithium metasilicate at 700 K. Characteristic times to separate different time regions are shown. The value $\tau_0$ of the Coupling model [7] is also shown. Reprinted with permission from Habasaki, J., Ngai, K. L., Hiwatari, Y. (2002), Molecular dynamics study of cage decay, near constant loss, and crossover to cooperative ion hopping in lithium metasilicate, *Phys. Rev. E*, 66, 021205, pp. 1–11. Copyright (2002) by the American Physical Society.

Based on this relation, characteristic times of MSD can explain why the different regions are found in conductivity spectra. The events in each region can be examined further in MD as discussed later.

The Haven ratio in Eq. (5.5) can be experimentally obtained from the ratio of tracer diffusion and conductivity. It can also be represented by the ratio

$$\sum_i < \mathbf{v}_i(0) \bullet \mathbf{v}_i(t) > / < \sum_i \mathbf{v}_i(0) \bullet \sum_j \mathbf{v}_j(t) > \tag{5.6}$$

as shown in the linear-response theory by Kubo [28]. The Haven ratio is regarded as an indicator not only of the geometrical correlations but also of the cooperativity in the ionic motions because it includes the cross term of ions *i* and *j*.

Needless to say, this expression can be used to calculate the ratio in MD in principle. However, one should note that it is not easy to obtain the Haven ratio in good statistics from MD data especially at low temperatures, because of the cooperative character of the denominator and existence of rare events. The same is true for conductivity itself, although several attempts have been done so far.

Even a MSD for the self-motion tends to show large fluctuations due to the contribution of rare events with cooperative nature. If the calculation time is not long enough, rare events will not be assessed. From the long run, many time series can be taken for analyses, while it may not work well to observe rare events by many independent short runs, because of an extremely large number of runs needed for this purpose [6].

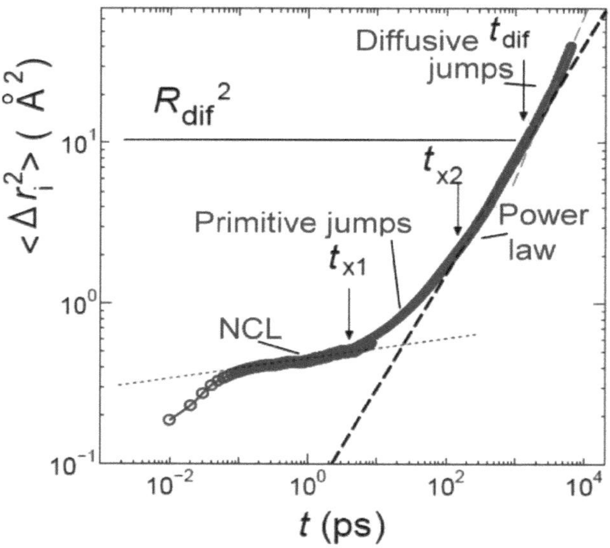


**Figure 5.3** Characteristic times found in MSD of Li ions to separate different functional forms. This example was taken for lithium disilicate glass at 800 K. Dashed and dotted lines (fitted to power laws) for Li ions at each temperature are for NCL and power law regions, respectively.

In Fig. 5.3, a schematic description of the MSD with characteristic times of Li ions in glassy $Li_2Si_2O_5$ systems is shown for the data at 800 K. It is separated by characteristic times, $t_{x1}$, $t_{x2}$, and $t_{dif}$. Comparable time regions as in lithium metasilicate are found in lithium disilicate, $Li_2Si_2O_5$, in the silica-rich region, and therefore the same assignments are applicable. In this figure, a concise explanation of each region and the relation between time scale and length scales are also depicted. Here the distance $R_{dif}$ corresponds to the mean distance among Li ion sites.

In silica-rich systems, the Haven ratio is known to be only slightly smaller than 1, and therefore changes in diffusive motion are representative of those in the conductivity.

By using MD, an examination of a wide region can be done relatively inexpensively. If one would like to examine the relaxation phenomena experimentally, usually combinations of several expensive apparatuses will be required.

## 5.1.2 Events in Each Characteristic Time Region

Each time-region in these MSDs can be characterized as follows by using van Hove functions, fractal dimension analyses [20], intermediate scattering functions and cage decay functions [1, 2, 10] for lithium silicate systems with several compositions [6]. It is worth to mention that quite similar behaviors are found in many systems, including non-ionic systems [6, 7].

Four distinct time regions for MSD of Li ions are found and time developing of it corresponds to the processes of ionic motion, from the localized motion within a cage (I) to final diffusive regimes (IV) through the regions with first jump motion of some ions (II), and power law regions (III), where bifurcated motions of fast and slow ions are mixing.

Region I; NCL region: After the initial overshoot of the ballistic motion corresponding to the boson peak, a nearly flat region assigned to the NCL is found, where the MSD increases with time as $t^{\alpha}$ with $\alpha \approx 0.07$. In this time region, ions are still caged by surrounding oxygen atoms and further shells of ions and atoms.

This behavior corresponds to the frequency dependence of the real part of the conductivity $\sigma'(\omega) \propto \omega^{1-\alpha}$ and the imaginary part of dielectric susceptibility, $\varepsilon''(\omega) \propto \omega^{-\alpha}$.

Thus, the time scale of $t_{x1}$ is related to the waiting time of jump motions.

Region II; Primitive jump region: Initial jumps of several ions are found in this region, where the MSD rises more rapidly than $t^{\alpha}$ of the regime (I). In this intermediate time regime of about $t_{x1} < t < t_{x2}$, MSD is nearly proportional to time and hence regarded as a quasi-diffusive regime. This statement can be easily confirmed by a linear scale plot.

It is interesting to note that the jump motion behaves as independent jumps here because the observation of squared displacement is usually done along with the direction of the first jump. After that, it is modified by angles among successive jumps. The jump can be a part of the successive jumps. In this sense, the difference between the first jump and the following jumps depends on when one starts the observation [19].

Region III; Power law region: Power law behaviors of MSD are found in the region $t_{x2} < t < t_{\text{dif}}$. In this region, MSD depends on $t^{\theta}$ with $\theta < 1$ and the small value of $\theta$ means the existence of slower dynamics compared with normal diffusion with the Gaussian type. Both fast and slow ions are found here and the slope $\theta$ is affected by the back-correlated motion of jumps considerably in the glass. The exponent $\theta$ is a function of the fractal dimension of the random walk [35], $d_{\text{w}}$, which is an index to represent the complexity of trajectories. For the mean behavior of ions, $\theta$ is expressed by Eq. (5.7) [36].

$$\theta = 2/d_{\text{w}}, \tag{5.7}$$

Region IV: Diffusive region: Diffusion coefficient of ions can be calculated after $t_{\text{dif}}$, where MSD reaches the square of the typical neighboring distance of ion sites and the MSD shows a linear $t$ dependence in that region. Thus, the time scale of $t_{\text{dif}}$ determines the elementary process of diffusive motion modified by correlations among successive motions. One should note that the activation energy calculated from the diffusion coefficients is not for elemental jumps but the mean behaviors of both fast and slow ions. With decreasing temperature, each time-region becomes longer and longer. Namely, the diffusive regime is found after 2 ns at 800 K for Li ions, while it is found after ~15 ns at 600 K. The validity of these assignments can be confirmed by the functional forms and van Hove functions as well as observed values of $<r_i^2>$. The value at $t_{dif}$ corresponds to the squared value of the neighboring ion sites. It is noteworthy that this behavior is for the average of slow and fast ions of extremely heterogeneous and non-Gaussian dynamics for single alkali silicate and related systems.

In Chapter 6, the characteristics of each time-region in lithium disilicate will be compared with the motion in the modified system by introducing pores.

In the following sections, details of ionic motions will be examined from caged ion dynamics to diffusive dynamics step by step.

Many examples are taken from works for both lithium disilicate and lithium metasilicate systems and related ones. Of course, there are some differences among systems, but the following discussions hold for these cases with some shifts in time and temperature regions.

## 5.1.3 Caged Ion Dynamics in NCL Region

### 5.1.3.1 Localized motions of ions in NCL regions

The characteristics of the caged ion dynamics are clearer at lower temperatures. The motion within a cage tends to continue for a long time. This situation is clear when individual motions of ions are examined. Squared displacements of localized motions of Li ions in NCL regions for arbitrarily chosen 6 ions in lithium disilicate are shown in Fig. 5.4 [10]. Each ion has a different color in each figure. The impression gathered from the figure may be different depending on the time intervals of data points used. Figures 5.4a,b show motions at 300 K during the time scale of NCL for the same ions but at different time intervals. In both figures, the active and quiet states intermittently appeared for the two ions shown in red and blue. Other ions are more localized. Strong heterogeneity of the motion is also observed at 500 K during the time scale of NCL (10 ps) in Fig. 5.4c. Here strong heterogeneity means the coexistence of ions with large displacements and immobile ions, although almost ions are still within cages. There are some remarkable characteristics in the motion of ions in the caging region.

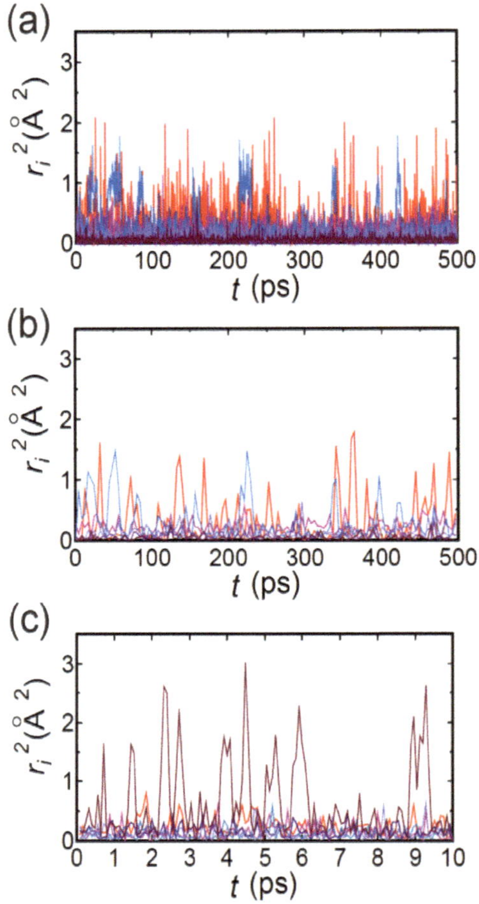


**Figure 5.4** Squared displacements of arbitrarily chosen 6 Li ions during the time scale of the NCL region. (a) At 300 K, (1250 data points for each ion). (b) 300 K (125 data points for each ion.) (c) 500 K (125 data points for each ion). Large displacements are found intermittently. The motion is not a harmonic vibration. Some ions show large squared displacements, while other ions are strongly localized. Adapted with permission from Habasaki, J., Ngai, K. L. (2017), Nearly constant loss (NCL) in lithium metasilicate glass at low temperatures—anisotropic and dynamical caging from molecular dynamics simulations, *J. Phys. Chem. C*, 121, pp. 13729–13737. Copyright (2017) American Chemical Society.

1. Ionics in the glassy state are dominated by jump motion of ions and it has intermittent nature related to chaos [37], where both laminae and turbulent states appeared at

different intervals. Similar intermittency is found in many systems including the behavior of nature such as rainfalls [38]. The motion is also sporadic [39, 40] so that rare events such as cooperative motions of several ions occur. Even though low probability, such cooperative and intermittent motions tend to show a remarkable contribution to the mean value of dynamics because of the long-range scale of motion with forward correlated motions. In the low-temperature regions, pre-jump motions contribute to MSD, while jump motions become rarer.

2. The motion of ions is rather deterministic as explained later. A large deviation from the harmonicity in the motion of ions in cages exists as shown in the following section and anharmonicity is suggested to be the cause of the NCL.

#### 5.1.3.2 Anharmonic and intermittent characters of caged ion dynamics in nearly constant loss region: shapes of the density profile of localized ions

Here we will examine what determines the motion of ions in cages.

The local motion of ions in the NCL region is affected by the positions of surrounding atoms and hence the motion is governed by the geometry of cages and their deformations. That is, the shape of the set of points in each potential well (site) in these figures is not necessarily spherical but is ellipsoid-like with some distortions as shown in Fig. 5.5, where an example of positions visited by Li ions at 300 K in the NCL regime is shown. Deviation from the spherical shape is observed and it was correlated well for the structure of cages formed by surrounding oxygen atoms [8, 41, 42]. As a natural consequence, the transformation of cages is governed by the geometric rules of coordination polyhedra.

Characteristics of the anharmonic motions are summarized as follows.

1. Within their cages, Li ions near the edge of the site tend to be sharply drawn back within the central part of the site. These motions correspond to the intermittent spike-like motions found in Fig. 5.4. Therefore, on a longer time scale than vibrations, the ion exhibited anharmonic motion. Jumps, including trial ones, are related to deterministic chaos as

shown by the phase-space plot. It is previously clarified for each ionic motion in lithium metasilicate and also in ionic liquid. [6, 43]. That is, anharmonic motions of ions in cages are followed by jumps having the deterministic nature.

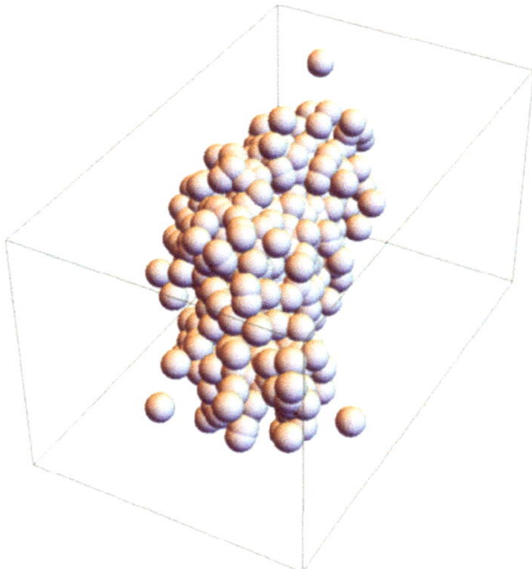

**Figure 5.5** An example of the three-dimensional plot of the positions of a Li ion during the MD run at 300 K, which exhibited the non-spherical shape. It means a contribution of the higher order terms of the multipole expansion. Positions are accumulated for 500 ps, that is, during the NCL region. The point is plotted at every 0.4 ps.

2. This non-spherical shape of density of ions means contributions of dipole and quadrupole of the ionic interactions, and therefore the charge–charge interaction is not enough to represent the motion within the cage. Then, the multipole expansion of the potential field will be useful to represent such situations. In other words, anharmonicity of the potential was considered as an origin of the NCL [6, 7, 10], because the contents of the potential in the NCL regime can be represented by the multipoles.
3. As shown in several works for the lithium silicates [10, 42, 43] as well as an ionic liquid [44, 45], the local field in the caging regime is dynamically fluctuating, where the large displacement of Li ions to the next site is accompanied with the changes in the coordination number of surrounding

oxygen atoms. A similar result is also observed in a colloidal system [46] in spite of the fact that the system is quite different from the ionic conductor examined here.

Lower than 500 K in lithium disilicate systems, it is not easy to observe the diffusive motions within ~ns order simulations. However, the comparison with the temperature dependence of dynamics obtained by MD and that of experimental values is successful [6, 47], and this fact means the caged ion motions observed in MD are comparable to experimental ones. The conductivity is still measurable down to ~300 K or below in experiments.

As shown in Chapters 6 and 7, the enhancement of the dynamics of Li ions is found by introducing pores to the system in the NVE conditions. The enhancement is found to be accompanied with the structural changes of cages (coordination polyhedra), which causes a shortening of the time scale of the caging region. Thus, the importance of the caged ion dynamics is found in such cases. In some theories or models in ionics or glass transition, the importance of the NCL region tends to be neglected, while it is pointed out in the Coupling Model by Ngai [3, 6].

As found in several properties in each region, ions escaping from ion sites show jump (hopping) motions to neighboring sites and after power law regions, finally they contribute to diffusive motions. Therefore, the role of motions in the NCL region is determining jump rates, and motions in cages are regarded as precursors of the following processes.

#### 5.1.3.3 Non-decaying part obtained from the Van Hove function found in the NCL region at low temperatures

The van Hove function can be used for further characterization of the time-regions found in MSD. The characteristics of the MSD in the NCL region at lower temperature regions are examined by the self-part of the van Hove function defined as follows [48, 49]:

$$G_s(\mathbf{r},t) = (1/N)\sum_{i=1}^{N}\left\langle \delta(\mathbf{r}_i(t) - \mathbf{r}_i(0) - \mathbf{r})\right\rangle \tag{5.8}$$

In Fig. 5.6, the self-part of the van Hove function at 300 K is shown, where $r \equiv |\mathbf{r}|$. At temperatures ~500 K and above, development of the second and further shells by jump motion is clear within a time scale of observation (~several ns), while at 300 K contribution at around $r$ = 3 Å is found only at the longest time. Therefore, at low temperatures, Li ions are almost trapped within original sites during the observation time. This time-region becomes considerably longer at low temperatures than those at higher temperatures.

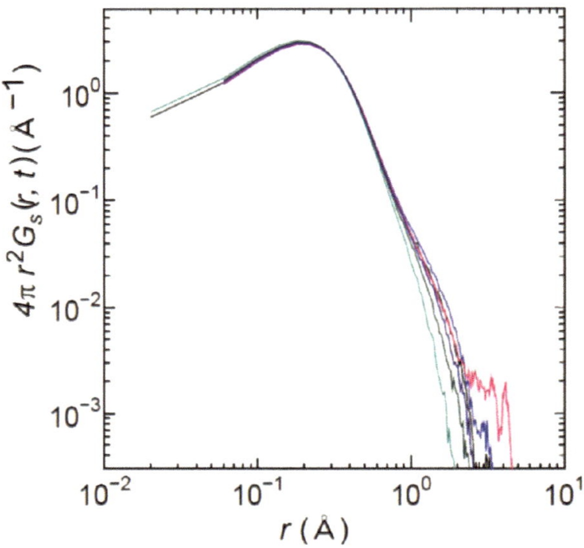


**Figure 5.6** Self-part of the van Hove function of Li ions in lithium disilicate at 300 K is shown, where $r \equiv |\mathbf{r}|$ in a log-log plot. Curves from inner to outer are for 0.8, 8, 39.2, 80.0, 560, and 1520 ps, respectively. Time development is small but is clear. Contribution at around 3 Å by the first jumps is found at the longest time in the plot.

Further details of the motion in the NCL region are discussed here for the lithium metasilicate system using both MSD and cage decay function. Another evidence of lengthening of the NCL region at low temperature will be shown as follows. In ref. [10], the temperature dependence of MSD of Li ions, including lower temperature data (at 100 K) and corresponding cage decay functions $N_{\text{cage}}(t)$ [1, 46] were examined. These results are shown in Fig. 5.7a,b.

In Fig. 5.7a, the temperature dependence of the MSD of Li ions is shown. As expected, with decreasing temperature, each time-region becomes longer and longer. At the low-temperature regions, ions are localized for a long time because of the long lifetime of the cage.

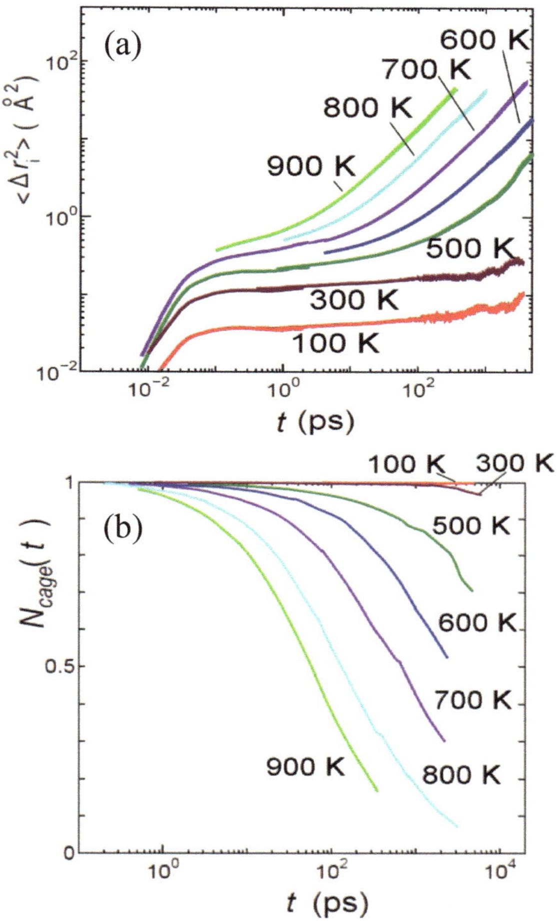


**Figure 5.7** (a) Mean squared displacement of Li ions in $Li_2SiO_3$ at 7 temperatures. (b) Corresponding cage decay function measured by the area under the first peak of the self-part of the van Hove function (normalized to 1 at the initial time) for the same temperatures. The function decays by the net jump to the neighbor or further shells. The data of the same temperature are represented by the same color. Adapted with permission from Habasaki, J., Ngai, K. L. (2017), Nearly constant loss (NCL) in lithium metasilicate glass at low temperatures—anisotropic and dynamical caging from molecular dynamics simulations, *J. Phys. Chem. C*, 121, pp. 13729–13737. Copyright (2017) American Chemical Society.

The cage decay functions are defined as follows:

$$N_{\text{cage}}(t) \equiv \int_{o}^{r_c} 4\pi r^2 G_s(r,t)dr. \tag{5.9}$$

Here, the summation is taken within a distance $r_c$ of the initial cage (less than 1.7 Å). The cut-off value corresponds to the first minimum of the self-part of the van Hove function when the clear second peak exists (500–800 K). The area under the curve of the first peak of the self-part of the van Hove function is proportional to the number of particles in cages. In this figure, the area is normalized to 1 at the initial time.

During the NCL region, essentially no ion can jump out of the cage, although in the case of ionically conducting glasses near room temperature, there is still a low probability for jumps.

As shown in Fig. 5.7b, $N_{\text{cage}}(t)$ at each low temperature does do not decay for a long time. At the lowest temperature, the NCL region has covered the entire observation time during the MD runs. It is worth to mention that the structural arrest observed in molecular dynamics often is concerned with the lengthened NCL (the caging region in non-ionic systems) and not necessarily mean the loss of diffusivity.

Therefore, careful treatment and serious consideration of this region found by molecular dynamics (MD) simulations in glass-forming materials are essential for a fundamental understanding of the glass transition.

For calculations of transport coefficients such as diffusion coefficient and corresponding DC conductivity, one needs to use the long-time region after $t_{\text{dif}}$, although it is not necessarily easy due to a limited simulation time up to ~ns in many cases. The covered time scale may be restricted by the real-time required to do such simulations.

### 5.1.4 Relation between the Power Law Behavior of MSD and Trajectories of Ions

In Fig. 5.8, positions of atoms and ions at 600 K during 8 ns run for lithium ions in lithium disilicate are shown. In this case, the time scale of the observation is longer than $t_{\text{dif}}$, even though both localized and diffusive motions along jump paths can be observed.

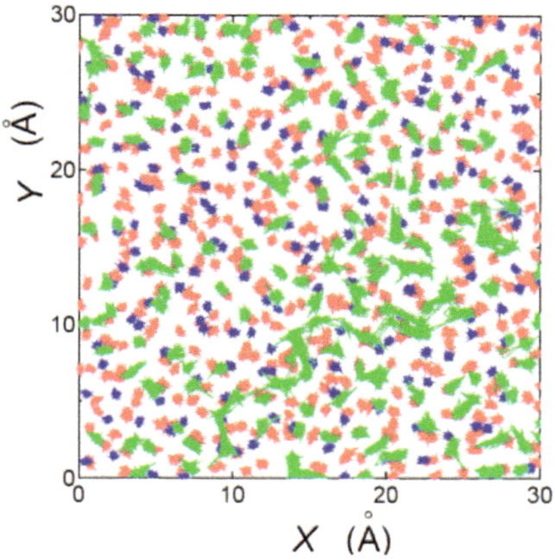


**Figure 5.8** Trajectories of Li ions in $Li_2Si_2O_5$ glass at 600 K for a slice of the system during 8 ns run. The motion is highly heterogeneous. That is, both localized jumps and diffusive jumps within ion channels are observed.

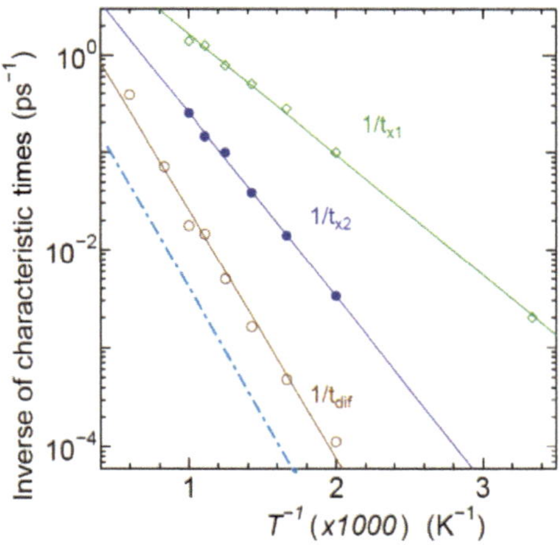


**Figure 5.9** Temperature dependence of characteristic times in MSD for Li ions in lithium metasilicate replotted from the data shown in [17]. The slope of diffusion coefficient is comparable to that for $1/t_{\text{dif}}$. The dotted dash line is for diffusion coefficient, which is arbitrarily shifted vertically for comparison. Larger changes in slopes are found at longer time scales. Activation energy of the $1/t_{\text{dif}}$ is comparable to the diffusive motion. Adapted with permission from Habasaki, J., Ngai, K. L. (2017), Nearly constant loss (NCL) in lithium metasilicate glass at low temperatures—anisotropic and dynamical caging from molecular dynamics simulations, *J. Phys. Chem. C*, 121, pp. 13729–13737. Copyright (2017) American Chemical Society.

Positions of Si and O atoms are essentially localized except for some fluctuation. The set of points for Li in the site is separated from the others but tends to show a pattern related to the jump paths observed at higher temperatures or at a longer time scale (see also Fig. 3.4 and Fig. 3.6 in Chapter 3 for examples of alignment and motions of ions). Each trajectory is the complex ones due to correlated motions, while in functions such as van Hove functions, net changes at distance *r* are measured. One can easily imagine that the localized motion within a cage becomes dominant at low temperatures. From such trajectories, the power-law region of MSD can be characterized using the fractal dimension analysis.

### 5.1.4.1 Cooperative and correlated jumps of ions

Deviations from the simple random walks are often found in MD simulations in limited time regions. The motion of ions in ionically conductive ions is highly cooperative as shown in early molecular dynamics works in the 1990s for $Ag^+$ ions [50] and $Li^+$ ions [51]. Ions tend to show forward correlated motions with long length scales. For the former, it was attributed to Frenkel-type defects. For the latter, it was pointed out that the distribution of displacements has an inverse power law tail (with an exponential truncation) so that the non-Gaussian dynamics related to this distribution are characterized [51] as Lévy flight dynamics [52]. In such motions, cooperative jumps play important roles. Similar dynamics are also known for non-ionic materials even for the MD simulations of the soft-core system without attractive terms [53].

In some of the early works, cooperative jumps were regarded to be very rare events, so that their contribution was negligible. This is probably because of a short time scale covered by simulations, which can be addressed by molecular dynamics due to the limited power of computers. However, later, it was pointed out that such motion should be the main mechanism in lithium metasilicate glass [51].

Similar cooperative jumps of several units have been observed in colloids [46], by confocal microscopy. This is a useful method for three-dimensional imaging and characterization of relaxation in colloidal systems as reviewed in [54].

Recently, cooperative jumps of ions seem to be "re-discovered" in several fast conductors [55–58]. It is worth to note that the existence of cooperative motions is confirmed in the works using in *ab initio* MD as well as classical MD and it means that the phenomena are essentially explained by the classical dynamics.

#### 5.1.4.2 Power law region and coexistence of fast and slow ions

Before the diffusive region region of MSD is reached, the power-law region is found between $t_{x2}$ and $t_{\mathrm{dif}}$. The power-law exponent in this region is correlated well with the complexity of the motion of ions with backward and forward correlated jumps. To characterize the power law exponent in this region, fractal dimension of random walk, $d_{\mathrm{w}}$, can be used (see Chapter 4). The value is a measure of the complexity of the motion of ions (or other units). The dimension defined by the following equation can be easily obtained from the trajectories of ions in MD:

$$N = AL^{-d_{\mathrm{w}}}, \tag{5.10}$$

where $L$ is a length of the divider and $N_T$ is the number of procedures using a divider to cover the trajectories. Thus, the value can be determined from the slope of the $N$ against $L$ in a double logarithmic plot.

For the free random walk, one can expect that $d_{\mathrm{w}} = 2$. The values of $d_{\mathrm{w}}$ larger than 3 mean that the trajectories are folded many times by the backward correlation and therefore the motion is localized and slowed down. On the other hand, a strong forward correlation reduces the values of $d_{\mathrm{w}}$.

As known in the theory of fractal for the walks of particles, the power law dependence $\sim t^{\theta}$ of the MSD, is characterized by $d_{\mathrm{w}}$ from the relation (5.10) [4], when the geometrical term governs the slope and if the motion is a simple walk. For example, the power law exponent of MSD was shown to be comparable to the $2/d_{\mathrm{w}}$ along with the composition dependence of the mixed alkali effect in ion dynamics, while in single alkali systems, one needs to consider contributions of motions with different length scales.

Thus, the analysis of the fractal dimension of the random walks is useful to examine the dynamical changes in the modification of the system.

### 5.1.5 Diffusive Region of MSD and Exchange of Fast and Slow Motion in Longer Time Scales

From the slope after $t_{\text{dif}}$, the diffusion coefficient, $D$, of ions can be determined by the following Einstein equation [59]:

$$D = \frac{1}{6}\lim_{t\to\infty}\frac{d}{dt} < [\mathbf{r}_i(t) - \mathbf{r}_i(0)]^2 >. \qquad (5.11)$$

where $D$ is the diffusion coefficient and $\mathbf{r}_i$ is the positional vector.

This relation holds when the path dynamically formed is percolated when $t_{\text{dif}}$ is attained. Therefore, the calculation of this coefficient should be done after $t_{\text{dif}}$.

Sometimes, the time region between $t_{x1}$ and $t_{x2}$ might be regarded as the diffusive region because of proportionality to time. This situation can be checked by the value of mean-squared displacement during the run. The $R_{\text{dif}}^2$ value at $t_{\text{dif}}$ is ~10 Å$^2$ in the case of lithium silicates and related systems and the smaller value often means that the time scale is too short to calculate diffusion coefficients (and corresponding conductivities).

Caution is required especially for the treatment of dynamics in complex systems. If the structures have domains with different sizes, the diffusive region of MSD may not be attained easily.

Furthermore, the dynamics of ions in ionically conducting material, as well as motions of molecules and/or atoms near the glass transition temperatures, are highly heterogeneous. Namely, some ions are localized for a long time, while some other ions tend to show accelerated motions with cooperativity as shown in the following section. To obtain the slope of MSD for calculation of diffusion coefficient, time windows to average the motions should be wide enough. Many initial times will be required to obtain the mean value of heterogeneous jump motions [6].

### 5.1.6 Temperature Dependence of Characteristic Times

The next step necessary to learn the complex ionics is to represent the trend found in the MSD by more quantitative manners. As shown in previous sections, the NCL region of MSD ends at the first characteristic time $t_{x1}$, which corresponds to the onset of the jump motion of ions or particles to neighboring sites. This time-region is followed by the start of another power-law dependence of MSD after $t_{x2}$, and finally, the long-term diffusive region starting after $t_{\text{dif}}$. It is useful to examine how these time-regions behave with the changes in temperature to learn what changes of motion occur. Temperature dependences of characteristics times, as well as diffusive motions, are analyzed to give a detailed description of the dynamics of the ions. Results of the inverse of the characteristic time of Li ions in the lithium disilicate system are shown in Fig. 5.9 as a function the inverse of temperature. As already mentioned, with decreasing temperature (or increasing pressure), each time-region is expected to be longer and longer. The absolute value of the slope for $1/t_{\text{dif}}$ is larger than that for $1/t_{x2}$. The value for $1/t_{x1}$ is the smallest but it also covers wide time scales. The slope for $1/t_{\text{dif}}$ is comparable to that for the diffusion coefficient, *D*.

In other words, the time scale of $t_{\text{dif}}$ determines the effective elemental process (as a mean behavior of fast and slow ions) for diffusivity. If the NCL region becomes long enough to cover the entire region of observation time, the MSD becomes flatter and the intermediate scattering function and/or other related correlation functions decay little. This situation can be well represented by the temperature dependence of the inverse of characteristic times.

In the longer time region, time development of ionic motion is modified by the correlation among successive jumps. The strong back correlation probability at lower temperatures contributes to the slowing down of the dynamics [6]. Characteristic times found in MSD are useful to judge what kind of motion is modified by the perturbation of the systems. The effect of introducing pores will be discussed in Chapter 6 by using these characteristic times.

## 5.2 Dynamic Heterogeneity: Coexistence of Slow and Fast Ions

Here, the term "dynamic heterogeneity" is used to imply the coexistence of both fast and slow diffusive dynamics of particles (ions, molecules). In this section, dynamic heterogeneity found in ionically conducting glasses is discussed, although the phenomena are common with other non-ionic glass-forming materials. Often, heterogeneity is regarded as the important properties to characterize glass transitions.

The concept is not necessarily the same as the fast and slow relaxations, because fast relaxation is usually accompanied with the local and/or rotational motions. One should keep in mind that the beginning of the researches of dynamic heterogeneity, the same or similar concepts were sometimes described by different terminology.

Self-part of the van Hove functions at 1673 K (in the molten state) and 800 K (near the glass transition temperature) for Li ions in lithium metasilicate, $Li_2SiO_3$ [60] are shown in Fig. 5.10.

In the molten state at 1673 K, the function is well represented by the Gaussian-type diffusive motion, while the large deviation is observed at lower temperatures.

In the power law region of MSD, the distribution of the displacement or corresponding behavior in the self-part of the van Hove functions shows an inverse-power law tail of the following form for the large $X$ (=$d|r|/\mathrm{d}t$ or $r$) region,

$$P(X) = \frac{1}{|X|^{\alpha+1}}, \tag{5.12}$$

with an exponential tail. This part of the van Hove functions is related to the Lévy (stable alpha) distribution (when $\alpha < 2$) as found in molten and glassy lithium metasilicate [51].

The broadening of the peak found in a short-time region represents the motion within a cage and the fluctuations of the cage itself, while the decreased area of the first peak represents the motion to the next shell. With the decay of the first peak, the tail part of the function develops with elapse of time. The long tail with large $r$ is related to the long-ranged motion as found in the trajectory, i.e., by fast ions with successive jump-like motions.

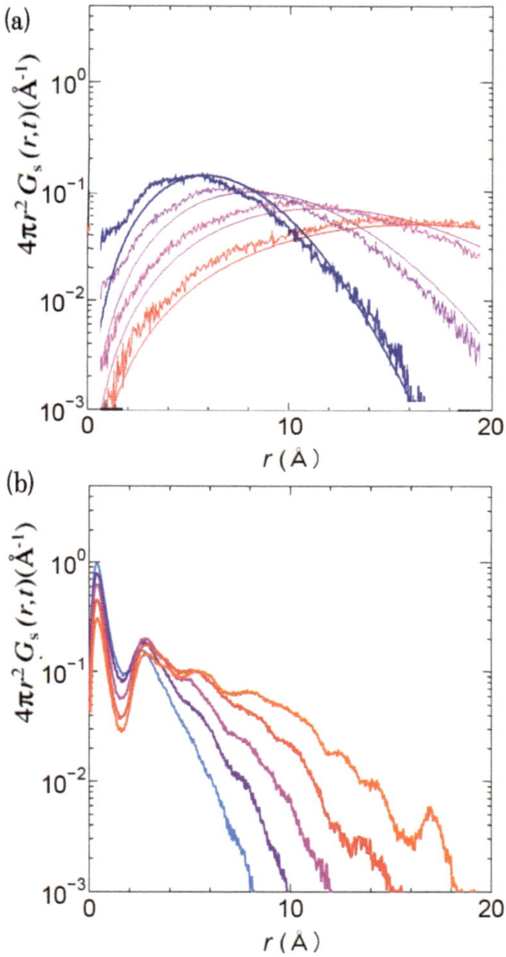


**Figure 5.10** (a) Self-part of the van Hove function of Li ions in lithium metasilicate at 1673 K (for $t$ = 16, 32, 64 and 128 ps from left to right in the curves of long ranged region). Smooth curves are for the Gaussian-type distribution having the same diffusion coefficient as the observed function. Same colors are used when the time is the same. Deviation from Gaussian form is observed only at short-time region. (b) Self-part of the van Hove function of Li ions in lithium metasilicate at 800 K (for $t$ = 50, 100, 200, 400, 800 ps from left to right in the in the curves of long ranged region) in lithium metasilicate, respectively. Both non-decaying part and jumps to the long length coexist. Reprinted with permission from Habasaki, J. (2007), On the nature of the heterogeneous dynamics of ions in ionic conducting glasses, *J. Non-Cryst. Solids*, 353, pp. 3956–3968. Copyright (2007), Elsevier.

As already mentioned, the fractal dimension of the random walk with $d_w = 2$ corresponds to the free random walk (Brownian diffusion). Fast ions with accelerated motion show smaller $d_w$ value ($d_w < 2$), which means the trajectories are more linear like and this also means the acceleration is achieved by the geometrical correlation among successive jumps.

Localization of particles (ions, molecules) is governed by the complexity of paths and motions of them. Localized motion tends to have larger $d_w$ ($> 2$) values, which means the trajectories are folded many times. On the path with small dimensionality, such motion can be understood as "fracton" [61], coming from the complicated motion within the path with small dimensionality. The concept of "fracton" can be explained as follows.

When the motion of ions is examined along the trajectory, the numbers of sites, $V(N)$, visited during the $N$ steps is represented by

$$V(N) \sim R_{\mathrm{df}} \sim Nd_s/2, \tag{5.13}$$

where $d_f$ is the density dimension of the cluster formed by the visited sites and $R$ is a linear size of it. Here, "fracton" dimension, $d_s$, is defined by

$$d_s = 2d_f/d_w. \tag{5.14}$$

That is, the vibrational state on the fractal structure depends on both the fractal dimension of the visited sites and the walk.

Then we consider a back-correlation probability, $<P_0(t)>$, of finding a walker. The walker is located at an origin initially at time $t_0$ and back to the initial position at time $t$.

The probability is proportional to an inverse of the numbers of the site, which is a function of the fracton dimension.

$$<P_0(N)> \sim [V(N)]^{-1}. \tag{5.15}$$

See ref. [7] for more details of dynamic heterogeneity in lithium silicates and ionic liquids.

### 5.2.1 Heterogeneity Found in Individual Ionic Motions

One fundamental question for the phenomena is if the fast and slow ions are part of the one distribution or mixture of different distributions. At least in the single alkali silicates, they can be regarded as parts of the one distribution (Lévy distribution with a truncation) [7, 51, 62] if the motions are averaged adequately, because of exchanges among fast and slow motions exist.

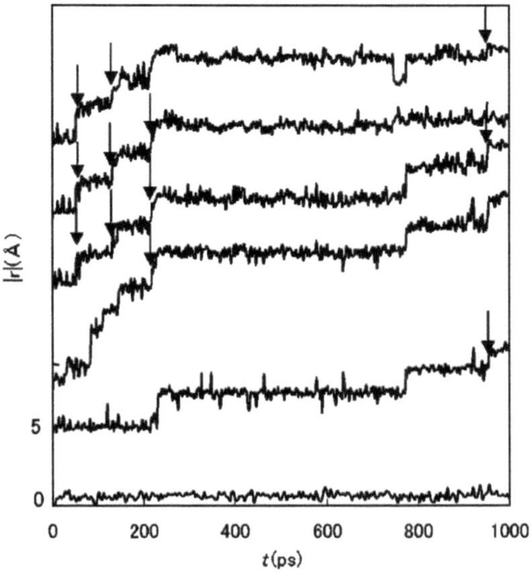


**Figure 5.11** Squared displacement of arbitrarily chosen 6 ions in lithium metasilicate at 700 K. Initial values are arbitrarily shifted to reduce overlaps. The lowest one is arrested during 1 ns run. Several ions cooperatively jump to the nearly same direction, as shown by arrows. Some localized jumps are also shown. Reprinted with permission from Habasaki, J., Hiwatari, Y. (2002), Dynamical fluctuations in ion conducting glasses: Slow and fast components in lithium metasilicate, *Phys. Rev. E*, 65, 021604, pp. 1–8. Copyright (2002) by the American Physical Society.

To understand the characteristics of dynamic heterogeneity, motions of ions including individual ones have been examined in several MD works [51, 63–67]. In Fig. 5.11, squared displacement of several arbitrarily chosen ions is shown. Both fast ions showing forward correlated jumps and slow ions showing a long waiting time of jumps and/or localized jumps are observed. Within ~1 ns, they tend to keep their characteristics; however, exchanges among these behaviors are observed. Trajectories of 7 ions with large mobility found in lithium metasilicate at 700 K are shown in Fig. 5.12. These ions tend to move together and show common paths. Such kind of paths (ion channels) formed by $SiO_4$ networks can change in position gradually at high temperature [68], especially at high alkali content glasses, it can also change due to network tends to be cut because "free oxygens" exist in the system [69–71].

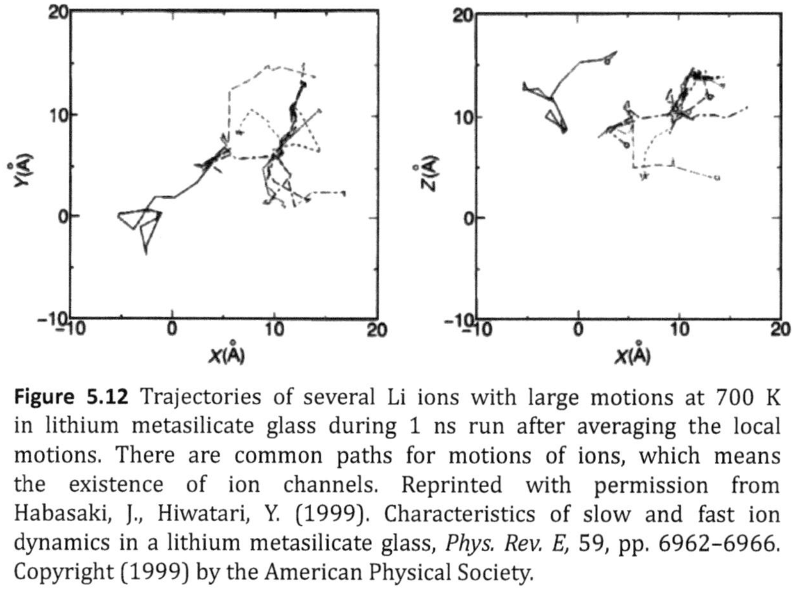


**Figure 5.12** Trajectories of several Li ions with large motions at 700 K in lithium metasilicate glass during 1 ns run after averaging the local motions. There are common paths for motions of ions, which means the existence of ion channels. Reprinted with permission from Habasaki, J., Hiwatari, Y. (1999). Characteristics of slow and fast ion dynamics in a lithium metasilicate glass, *Phys. Rev. E*, 59, pp. 6962–6966. Copyright (1999) by the American Physical Society.

That is, motions of ions are highly cooperative and dominated by the accelerated motions of a small number of ions.

### 5.2.2 Fractal Dimension Analyses of Jump Paths and Walks

The changes of the characteristic times in the temperature dependence have been remarkable at longer time scales. To explain the slowing down of the dynamics at longer time scales more quantitatively, the description of jump paths and walks by fractal dimension analyses is useful. It describes how the trajectory of ions spread over the space through the jump motions.

The fractal dimension of the path can be calculated in the cumulative number of the same kinds of ions within $r$ when one ionic species is concerned. From the slope of the power law region, one can estimate the dimension for the possible connections of the path during a certain time of MD runs.

Multifractal analyses of density profiles of ions [68] are also useful to judge the percolation of paths as shown in Fig. 4.5 in Chapter 4. This analysis was applied for the accumulated position of ions during a long-time run so that the result can cover both the localized motions and diffusive motions. In our case, several times longer time scales than $t_{\mathrm{dif}}$ are necessary for accumulation to judge the percolation of the path [21].

### 5.2.3 Relation between Temporal and Spatial Terms in Dynamics

The dimension of the path is not the same as that for the motions of ions. One can easily imagine a walker coming and going with hanging around on the way, which requires a long time to reach the destination. This means that the structure of the pathway alone cannot determine the behavior of the walker. This example also shows that how geometrical correlations with strong back correlation result in the temporal slowing down of the motion. This relation will be useful to consider the mechanism of the enhanced dynamics as well.

That is, analysis along the trajectory of the walker is required. The analysis of the topological nature of paths and motions is useful to understand dynamics [63, 64] and fractal dimension analysis of the random walk of trajectories of ions is applied successfully to understand the mixed alkali effect in mixed alkali silicates [21, 24]. Such an analysis is also useful to characterize the fast and slow components of diffusive motions, that is, dynamic heterogeneity [72]; that is, analysis of the coexistence of two different regions can be done as follows.

The fractal dimension of the random walks [35], that is the complexity of the trajectory obtained by simulation during the certain run time $t_{run}$, can be measured by using a divider method. We examined all the trajectories of the ions, obtained during long runs [43]. (For example, $t_{run}$ = 4 ns at 700 K for $x$ = 0, which is about 8 times longer than $t_{dif}$.) From the trajectories, we determine the fractal dimension of random walk, $d_w$, defined by

$$N = A'L^{-d_w}, \tag{5.16}$$

#### 5.2.2.1 Heterogeneous dynamics and mixing of them in longer time scales

The $d_w/2$ value corresponds to the power law exponent $\theta$ when the system is monofractal. A similar situation might be found in the fractal time random walk [73–75] when we used the effective localization time below $t_{dif}$. In that theory, waiting time distribution with a long tail and change in the distribution can be an origin of the dynamic slowing down. The role of back correlation for slowing down of the dynamics discussed in the previous subsection might be mathematically similar to that caused by the waiting time distribution and sometimes they are not distinguished from each other. However, particles with localized motions are not the same as immobile particles. That is, spatial term and the temporal term can be distinguished in the molecular dynamics if necessary.

First, we need to discuss which term is the origin of the power law of MSD. It can be clarified we have separated the contribution of temporal and spatial terms in MSD [7, 76–78]. In the case of lithium metasilicate glasses, it was clarified that the power law of MSD for motions of Li ions is due to the back-correlation and not by the waiting time distribution. It was found that the change of the temporal term can contribute to the slowing down of the dynamics through the smaller jump rate but not to the exponent of the power law directly.

There are several works reported for the role of the fractal structures on random walks [79–84] related to the slowing down of the dynamics. The importance of the fractal geometry has been pointed out [82] and that on cluster-cluster aggregates and multicenter diffusion limited aggregates [83] have been discussed based on the trajectories of particles obtained by using Monte Carlo simulations.

The scenario for mono-fractal behavior of ions does not necessarily hold in single alkali silicate systems [43] as well as ionic liquids [72]. In the fractal dimension analyses of trajectories in lithium metasilicate, the slope of $N$ against $L$ plot was found to change at around 3 Å at lower temperatures. Since two regions of slopes are found, the $d_w$ for a shorter length scale (<3 Å) is named $d_{w1}$ and a longer length scale ($3 < L < 10$) is named $d_{w2}$. Thus, the walks of Li ions in the lithium metasilicate have a multifractal nature. On the other hand, the difference between $2/d_{w1}$ and $2/d_{w2}$ is negligibly small for the mixtures as found in lithium-potassium metasilicate. It is also clear that $2/d_w$ value of trajectories with the geometrical origin is responsible for $\theta$ in the power law region of MSD, although generally, the fractal dimension of the random walk can include the temporal factor.

The role of the spatial term can be proved by the fractal structure of jumps related to the wavenumber dependence or by several works. Our definition of fast and slow ions is related to the power law distribution of length scales, which is related to the wavenumber dependence of the $\tau$ in $F_s(k, t)$.

In refs. [1] and [84], $F_s(k, t)_s$ for wide $k$ range for this system were examined.

$$F_s(k,t) = \langle \sum_{j=1}^{N} \exp\{i\mathbf{k}\cdot(\mathbf{r}_j(t) - \mathbf{r}_j(0))\}\rangle / N \tag{5.17}$$

The $F_s(k, t)$ for intermediate wavenumber (at around $k = 2\pi/2 \sim 2\pi/3$) is concerned with the back motion to the site occupied at $t = 0$. The back correlation is governed by the fracton dimension rather than the fractal dimension of the random walks. The effect of fast ions is larger in the small wavenumber for MSD due to squared distance. The fast ions contribute to the fast decay of $F_s(k, t)$ at small wavenumber as well. For the discussion of dynamic heterogeneity, thus the wavenumber dependence of it is useful.

To distinguish the role of the geometrical factor and the temporal factor is interesting for describing the dynamics of ionics as well as in examining glass-forming materials.

After $t_{dif}$, the distinction of fast and slow particles might be coarse-grained in observations but still exists. It may be found in the effective length scale and time scale of events on average, which are affected by the coexistence of fast and slow ions. In the diffusive regime, the geometrical correlation between successive jumps determines the effective length and time scales. These insights might be lost or substituted by the effects of the temporal term if the random walk scenarios without spatial terms are applied.

### 5.2.4 Exchanges between Fast and Slow Dynamics

It may be interesting to examine how fast and slow motions are mixing and change each other and how temporal and spatial terms are related to this. The data obtained by molecular dynamics simulations are adopted to examine further details of the dynamic heterogeneity.

Fast ions tend to have a short interval of jumps and show the forward correlated jumps, while slow ions tend to have longer intervals of jumps and/or strong back correlated jumps [64]. Therefore, it is interesting to examine the relation between temporal and spatial characters and its mixing of longer time scales than $t_{\text{dif}}$.

Examples of such results [63] will be shown here. In Fig. 5.13, the number of jumps, $N_{\text{i}}$, for 10 arbitrarily chosen Li ions in each time-region 0–1, 1–2, 2–3 and 3–4 ns of lithium metasilicate at 700 K is shown. Some ions keep their mobility in jump rate for a long time, while other ions tend to show changes of their mobility.

Resultant squared displacements of these ions are shown in Fig. 5.14.

For example, for Li-ions of No. 2 and No. 7, both the number of jumps and squared displacement in each time interval are small. In the case of Li-ion of No. 6, numbers of jumps are relatively large, while the squared displacement is small. This means the jumps are localized ones.

For the Li-ion of No. 1, the largest squared displacement is found in the third time region (2–3 ns); however, the jump rate here is not the largest. That is, a forward correlation among jumps contributes to the large squared displacement in this region.

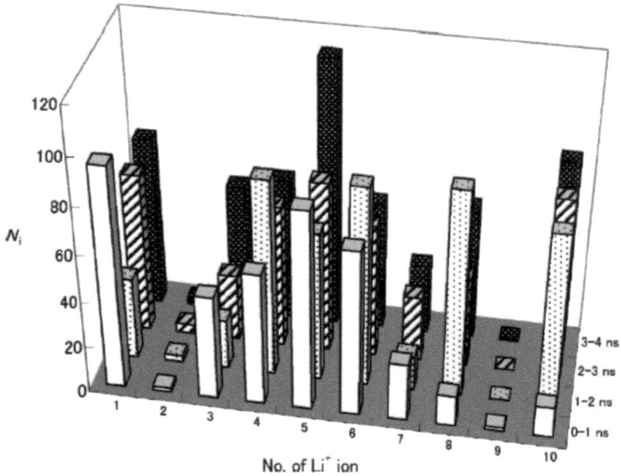


**Figure 5.13** Number of jumps, $N_i$, for 10 arbitrarily chosen Li-ions in each time-region of 0-1, 1-2, 2-3 and 3-4 ns of lithium metasilicate at 700 K. Reprinted with permission from Habasaki, J., Hiwatari, Y. (2002), Dynamical fluctuations in ion conducting glasses: Slow and fast components in lithium metasilicate, *Phys. Rev. E,* 65, 021604, pp. 1–8. Copyright (2002) by the American Physical Society.

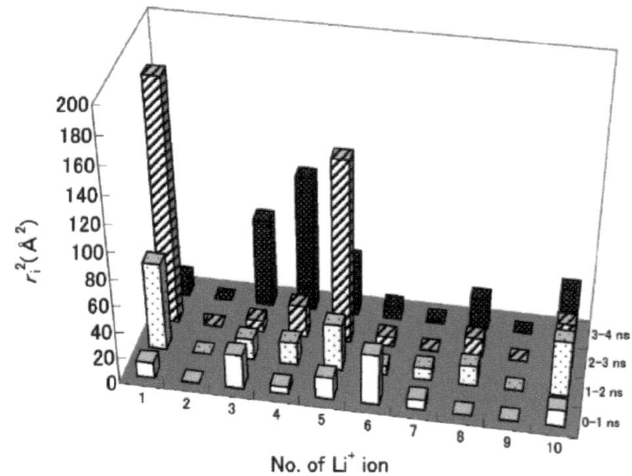


**Figure 5.14** Squared displacements $r_i^2$ for 10 arbitrarily chosen Li-ions in each time-region (the number of Li-ion corresponds to that in Fig. 5.13). Reprinted with permission from Habasaki, J., Hiwatari, Y., (2002), Dynamical fluctuations in ion conducting glasses: Slow and fast components in lithium metasilicate, *Phys. Rev. E,* 65, 021604, pp. 1–8. Copyright (2002) by the American Physical Society.

For Li-ions of No. 8 and No. 9, jump rates after 1 ns are larger than that for the 0–1 ns. However, squared displacements for these ions are kept small for 4 ns. Therefore, motions tend to be localized and the increase in the jump rate does not directly cause an increase in diffusivity. Displacements become large after 3 ns for the Li-ions of No. 3 and No. 4. A corresponding increase in the jump rate is found in the former case, while remarkable changes are not found in the latter case. In the former case, the change can be correlated with the increased jump rate. However, in the latter case, the change is mainly concerned with geometrical correlations rather than changing the jump rate. As found in these Figs. 5.13 and 5.14, changes in temporal characteristics related to the jump rate and spatial characteristics related to the geometrical correlations are not simple nor necessarily directly linked. Since heterogeneity in structures and/or dynamics is widely found in nanostructured materials such as catalysts [85], colloidal systems [86], it will be interesting to examine the relationship with the functions of systems and heterogeneity of systems.

## 5.3 Conclusion

To examine how dynamics of ions are affected by modifications such as introducing pores, some fundamental knowledge for the dynamics is required. In this chapter, it was shown that mean-squared displacements of ionic motion in lithium silicate glasses show several distinct time regions from caged ion dynamics to final diffusive motions. These time regions are the basis to consider how the modification of the system such as introducing pores affects the dynamics of systems and therefore useful to understand the changes in dynamics in nanostructured materials.

For example, the change in the short-time scale is concerned with the changes in the caging of ions in ionically conducting materials. The caged ion dynamics are well correlated to the structures of cages, where the concept of geometrical degree of freedom is useful. The caged ion dynamics region is assigned to the NCL known in experiments. One can expect that each time-region becomes longer with decreasing temperature. This lengthening discussed previously for the MSD in lithium metasilicate was confirmed by the simulations at lower temperatures. Furthermore, interpretations of the anharmonic nature of the ionic motion in the NCL region are given. The observed lengthening of the NCL region is comparable to that observed in non-ionic systems near the glass transition temperature.

Localized (caging) and diffusive regions for ionic motion are connected by the power law region, which can be characterized by the complexity of the trajectories.

Some of the further details of the characteristics of dynamic heterogeneity were discussed here.

## References


1. Habasaki, J., Hiwatari, Y. (1995), Origins of the two-step relaxation and the boson peak in an alkali silicate glass studied by molecular-dynamics simulation, *Phys. Rev. E*, 52, p. 2681.
2. Habasaki, J., Ngai, K. L., Hiwatari, Y. (2002), Molecular dynamics study of cage decay, near constant loss, and crossover to cooperative ion hopping in lithium metasilicate, *Phys. Rev. E*, 66, 021205(1–11).
3. Ngai, K. L. (2003), An extended coupling model description of the evolution of dynamics with time in supercooled liquids and ionic conductors, *J. Phys. Cond. Matt.*, 15, pp. S1107–S1126. DOI: 10.1088/0953-8984/15/11/332.
4. Kulkarni, A. R., Myer, F., Schnider, U., Lunkenheimer, P., Loidl, A. (1999), Ionic conductivity and relaxations in Li and (Li and K) fluorophosphate glasses, *Ceramic Trans.*, 92, p. 115.
5. Ngai, K. L., Leon, C. (2002), Cage decay, near constant loss, and crossover to cooperative ion motion in ionic conductors: Insight from experimental data, *Phys. Rev. B*, 66, 064308 (1–11).
6. Habasaki, J., León, C, Ngai, K. L. (2017), *Dynamics of Glassy, Crystalline and Liquid Ionic Conductors: Experiments, Theories, Simulations*, Springer International, Switzerland.
7. Ngai, K. L. (2011), *Relaxation and Diffusion in Complex Systems*. Springer, New York.
8. Habasaki, J. (2016), Molecular dynamics study of nano-porous materials—Enhancement of mobility of Li ions in lithium disilicate in NVE conditions, *J. Chem. Phys.*, 145, p. 204503.
9. Habasaki, J., Ngai, K. L. (2018), Heterogeneous dynamics in nano porous materials examined by molecular dynamics simulations—

effects of modification of caged ion dynamics, *J. Non-Cryst. Solid,* 498, pp. 364–371.

10. Habasaki, J., Ngai, K. L. (2017), Nearly constant loss (NCL) in lithium metasilicate glass at low temperatures—anisotropic and dynamical caging from molecular dynamics simulations, *J. Phys. Chem. C,* 121(25), pp. 13729–13737.
11. Bunde, A., Funke, K., Ingram, M. D. (1998), Ionic glasses: History and challenges, *Solid State Ionics,*105, pp. 1–13.
12. Dyre, J. C., Maass, P., Roling, B., Sidebottom, D. L. (2009), Fundamental questions relating to ion conduction in disordered solids, *Rep. Prog. Phys.,* 72, p. 4046501.
13. Angell, C. A. (1992), Mobile ions in amorphous solids, *Annu. Rev. Phys. Chem.,* 43, pp. 693–717.
14. Ngai, K. L. (1996), A review of critical experimental facts in electrical relaxation and ionic diffusion in ionically conducting glasses and melts, *J. Non-Cryst. Solids,* 203, pp. 232–245.
15. Day, D. (1976), Mixed alkali glasses—their properties and uses, *J. Non-Cryst. Solids,* 21(3) pp. 343–372.
16. Isard, J. O. (1969), The mixed alkali effect in glass, *J. Non-Cryst. Solids,* 1, pp. 235–261.
17. Ingram, M. D. (1987), Ionic conductivity in glass, *Phys. Chem. Glasses,* 28, pp. 215–234.
18. Ingram, M. D. (1994), The mixed alkali effect revisited—a new look at an old problem, *Ber. Glass Sci. Technol.,* 67, pp. 151–155.
19. Tomandl, G., Schaeffer, H. A. (1985), The mixed-alkali effect-A permanent Challenge, *J. Non-Cryst. Solids,* 73, pp. 179–196.
20. Bunde, A., Ingram, M. D., Maass, P. (1994), The dynamic structure model for ion transport in glasses, *J. Non-Cryst Solids,* 172–174, pp. 1222–1236.
21. Habasaki, J., Ngai, K. L. (2007), The mixed alkali effect in ionically conducting glasses revisited: A study by molecular dynamics simulation, *Phys. Chem. Chem. Phys.,* 9, pp. 4673–4689.
22. Balasubramanian, S., Rao, K. J. (1993), Preferential paths in alkali ion migration and the mixed alkali effect in silicate glasses, *J. Phys. Chem.,* 35, pp. 8835–8838.
23. Habasaki, J., Okada, I., Hiwatari, Y. (1995), MD study of mixed alkali effect in a lithium-potassium metasilicate glass, *J. Non-Cryst. Solids,* 183, pp. 12–21.

24. Habasaki, J., Okada, I., Hiwatari, Y. (1996), MD study of the mixed alkali effect in terms of the potential surface in the lithium-potassium metasilicate glass, *J. Non-Cryst. Solids,* 208, pp. 181–190.

25. Lammert, H., Heuer, A. (2005), Contributions to the mixed-alkali effect in molecular dynamics simulations of alkali silicate glasses, *Phys. Rev. B,* 72, 214202 (1–9).

26. Jonscher, A. K. (1977), The 'universal' dielectric response, *Nature*, 267, pp. 673–679.

27. Jonscher, A. K. (1983), Dielectric Relaxation in Solids (Chelsea, London).

28. Almond, D. P., West, A. R., Grant, R. J. (1982), Temperature dependence of the a.c. conductivity of Naβ-alumina, *Solid State Commun.* 44, pp. 1277–1280.

29. Moynihan, C. T. (1994), Analysis of electrical relaxation in glasses and melts with large concentrations of mobile ions, *J. Non. Cryst. Solids,* 172–174, pp. 1395–1407.

30. Moynihan, C. T. (1998), Description and analysis of electrical relaxation data for ionically conducting glasses and melts, *Solid State Ionics*, 105, pp. 175–183.

31. Ngai, K. L., Moynihan, C. T. (1998), The dynamics of mobile ions in ionically conducting glasses and other materials, *MRS Bull.*, 23, pp. 51–56.

32. Kubo, R. (1957), Statistical-mechanical theory of irreversible processes. I. General theory and simple applications to magnetic and conduction problems, *J. Phys. Soc. Jpn.*, 12, pp. 570–586.

33. Odagaki, T., Lax, M. (1981), Coherent-medium approximation in the stochastic transport theory of random media, *Phys. Rev. B,* 24, pp. 5284–5294.

34. Haven, Y., Verkerk, B. (1965), Diffusion and electrical conductivity of sodium ions in sodium silicate glasses, *Phys. Chem. Glasses,* 6, pp. 38–45.

35. Mandelbrot, B. (1967), How long is the coast of Britain? statistical self-similarity and fractional dimension, *Science*, 156, pp. 636–638.

36. Ben-Avraham, D., Havlin, S. (2000), *Diffusion and Reactions in Fractals and Disordered Systems*, Cambridge University Press, Cambridge.

37. Wang, X.-J. (1989), Statistical physics of temporal intermittency, *Phys. Rev. A,* 40, pp. 6647–6661.

38. Rigby, J. R., Porporato, A. (2010), Precipitation, dynamical intermittency, and sporadic randomness, *Adv. Water Resour.*, 33, pp. 923–932.

39. Wang, X.-J. (1995), Sporadic chaos in space-time dynamical processes, *Phys. Rev. E*, 52, pp. 1318–1324.

40. Wang, X.-J. (1992), Dynamical sporadicity and anomalous diffusion in the Lévy motion, *Phys. Rev. A*, 45, pp. 8407–8417.

41. Habasaki, J. (1991), Molecular dynamics study of microscopic mechanism of diffusion in $Li_2SiO_3$ system, *Z. Naturforsch.*, 46a, pp. 616–620.

42. Habasaki, J. (1990), Molecular dynamics study of the glass formation in $Li_2SiO_3$ system, *J. Habasaki Mol. Phys.*, 70, pp. 513–528.

43. Habasaki, J., Ngai, K. L., Hiwatari, Y. (2005), Time series analysis of ion dynamics in glassy ionic conductors obtained by a molecular dynamics simulation, *J. Chem. Phys.*, 122, 054507 (1–10), https://doi.org/10.1063/1.1824034.

44. Habasaki, J., Ngai, K. L. (2008), Heterogeneous dynamics of ionic liquids from molecular dynamics simulations, *J. Chem. Phys.*, 129, 194501 (1–16).

45. Habasaki, J., Ngai, K. L. (2015), Rigidity and soft percolation in the glass transition of an atomistic model of ionic liquid, 1-ethyl-3-methyl imidazolium nitrate, from molecular dynamics simulations—Existence of infinite overlapping networks in a fragile ionic liquid, *J. Chem. Phys.*, 142, 164501 (1–18).

46. Weeks, E. R., Crocker, J. C., Levitt, A. C., Schofield, A., Weitz, D. A. (2000), Three-dimensional direct imaging of structural relaxation near the colloidal glass transition, *Science*, 287, pp. 627–631.

47. Gonçalves, L. G. V., Rino, J. P. (2014), Diffusion mechanisms in lithium disilicate melt by molecular dynamics simulation, *J. Non. Cryst. Solids*, 402, pp. 91–95.

48. van Hove, L. (1954), Correlations in space and time and born approximation scattering in systems of interacting particles, *Phys. Rev.*, 95, pp. 249–262.

49. Glauber, R. J. (1955), Time-dependent displacement correlations and inelastic scattering by crystals, *Phys. Rev.*, 98, pp. 1692–1698 and references therein.

50. Kaneko, Y., Ueda, A. (1988), Molecular dynamics studies of superionic conductor $CaF_2$, *J. Phys. Soc. Jpn.*, 57(9), pp. 3064–3073. DOI: 10.1143/JPSJ.57.3064.

51. Habasaki, J., Okada, I., Hiwatari, Y. (1997), Fracton excitation and Lévy flight dynamics in silicate glasses, *J. Phys. Rev. B,* 55, p. 6309.

52. Klafter, J., Shlesinger, M. F., Zumofen, G. (1996), Beyond brownian motion, *Phys. Today*, p. 33, and references therein.

53. Miyagawa, H., Hiwatari, Y., Bernu, B., Hansen, J. P. (1988), Molecular dynamics study of binary soft-sphere mixtures: Jump motions of atoms in the glassy state, *J. Chem. Phys.,* 88, pp. 3879–3886.

54. Prasad, V., Semwogerere, D., Weeks, E. R. (2007), Confocal microscopy of colloids, *J. Phys. Condens. Matter,* 19, 113102 (25pp).

55. Deng, Y., Eames, C., Chotard, J.-N., Islam, M. S. (2015), Structural and mechanistic insights into fast lithium-ion conduction in $Li_4SiO_4$-$Li_3PO_4$, solid electrolytes, *J. Am. Chem. Soc.,* 137, pp. 9136–9145. DOI: 10.1021/jacs.5b04444.

56. Panchmatia, P. M., Armstrong, A. R., Bruce, P. G., Islam, M. S. (2014), Oxygen diffusion in yttria stabilised zirconia—experimental results and molecular dynamics calculations, *Phys. Chem. Chem. Phys.*, 16, pp. 21114–21118.

57. Nishino, S., Fujiwara, T., Yamasaki, H. (2014), Nanosecond quantum molecular dynamics simulations of the lithium superionic conductor $Li_{4-x}Ge_{1-x}P_xS_4$, *Phys. Rev. B,* 90, p. 024303.

58. Fujimura, K., Seko, A., Koyama, Y., Kuwabara, A., Kishida, I., Shitara, K., Fisher, C. A. J., Moriwake, H., Tanaka, I. (2013), Accelerated materials design of lithium superionic conductors based on first-principles calculations and machine learning algorithms, *Adv. Energy Mater.,* 3, pp. 980–985.

59. Einstein, A. (1905), Über die von der molekularkinetischen Theorie der Wärme geforderte Bewegung von in ruhenden Flüssigkeiten suspendierten Teilchen, *Annalen der Physik*, 322(8), pp. 549–560.

60. Habasaki, J. (2007), On the nature of the heterogeneous dynamics of ions in ionic conducting glasses, *J. Non-Cryst. Solids,* 353, pp. 3956–3968.

61. Alexander, S., Orbach, R. (1982), Density of states on fractals: «fractons», *J. Phys. (Paris) Lett.*, 43, pp. L625–L631.

62. Ben-Avraham, D., Havlin, S. (1983), Exact fractals with adjustable fractal and fracton dimensionalities, *J. Phys. A: Math. Gen.* 16, L559–L563. https://doi.org/10.1088/0305-4470/16/15/002.

63. Habasaki, J., Hiwatari, Y. (2002), Dynamical fluctuations in ion conducting glasses: Slow and fast components in lithium metasilicate, *Phys. Rev. E,* 65, 021604(1–8).

64. Habasaki, J., Hiwatari, Y. (1999), Characteristics of slow and fast ion dynamics in a lithium metasilicate glass, *Phys. Rev. E,* 59, pp. 6962–6966.

65. Banhatti, R. D., Heuer, A. (2001), Structure and dynamics of lithium silicate melts: Molecular dynamics simulations, *Phys. Chem. Chem. Phys.,* 3, pp. 5104–5108.

66. Varsamis, C-P. E., Vegiri, A., Kamitsos, E. I. (2002), Molecular dynamics investigation of lithium borate glasses: Local structure and ion dynamics, *Phys. Rev. B,* 65, p. 104203.

67. Kunow, M., Heuer, A. (2005), Coupling of ion and network dynamics in lithium silicate glasses: A computer study, *Phys. Chem. Chem. Phys.,* 7, pp. 2131–2137.

68. Habasaki, J., Ngai, K. L. (2005), Multifractal analysis of dynamic potential surface of ion-conducting materials, *J. Chem. Phys.,* 122, 214725 (1–10), https://doi.org/10.1063/1.1924691.

69. Nesbitt, H. W., Bancroft, G. M., Henderson, G. S., Ho, R., Dalby, K. N., Huang, Y., Yan, Z. (2011), Bridging, non-bridging and free ($O^{2-}$) oxygen in $Na_2O$-$SiO_2$ glasses: An X-ray photoelectron spectroscopic (XPS) and nuclear magnetic resonance (NMR) study, *J. Non-Cryst. Solids,* 357, pp. 170–180.

70. Nesbitt, H. W., Henderson, G. S., Bancroft, G. M., Sawyer, R., Secco, R. A. (2017), Bridging oxygen speciation and free oxygen ($O^{2-}$) in K-silicate glasses: Implications for spectroscopic studies and glass structure, *Chem. Geol.,* 461, pp. 13–22.

71. Stebbins, F. (2017), "Free" oxide ions in silicate melts: Thermodynamic considerations and probable effects of temperature, Chemical, *Geology*, 461, pp. 2–12.

72. Habasaki, J., Ngai, K. L. (2008), Molecular dynamics study of the dynamics near the glass transition in ionic liquids, *Anal. Sci.,* 24, pp. 1321–1327.

73. Montroll, E., Weiss, G. H. (1965), Random walks on lattices. II, *J. Math. Phys.,* 6, pp. 167–181.

74. Germano, G., Politi, M., Scalas, E., Schilling, R. L. (2009), Stochastic calculus for uncoupled continuous-time random walks, *Phys. Rev. E,* 79, p. 066102.

75. Meerschaert, M. M., Scalas, E. (2006), Coupled continuous time random walks in finance, *Phys. A: Stat. Mech. its Appl.,* 370(1), pp. 114–118.

76. Habasaki, J., Okada, I., Hiwatari, Y. (1998), Ion dynamics in pure and mixed alkali glasses-separation of the spatial and temporal aspects, *J. Phys. Soc. Jpn.*, 67, pp. 2012–2017.

77. Habasaki, J., Hiwatari, Y. (2002), Molecular dynamics study of single and mixed alkali metasilicates-spatial and temporal characterization of the dynamics in the supercooled liquid and glassy states, *J. Non-Cryst. Solids*, 307–310C, pp. 930–938.

78. Habasaki, J., Ngai, K. L. (2008), Refinements in the characterization of the heterogeneous dynamics of Li ions in lithium metasilicate, *J. Chem. Phys.*, 129, 034503 (1–11).

79. Höfling, F., Franosch, T. (2013), Anomalous transport in the crowded world of biological cells, *Rep. Prog. Phys.*, 76(4), p. 046602. doi: 10.1088/0034-4885/76/4/046602.

80. Yonezawa, F., Fujiwara, S., Gomi, S. (1996), Model of anomalous relaxation in supercooled liquids: Random walk in fractal space and time, *J. Non-Crystal. Solids*, 205–207, pp. 884–887.

81. Olveczky, B. P., Verkman, A. S. (1998), Monte Carlo analysis of obstructed diffusion in three dimensions: Application to molecular diffusion in organelles, *Biophys. J.*, 74, pp. 2722–2730.

82. Saxton, M. J. (1994), Anomalous diffusion due to obstacles: A Monte Carlo study. *Biophys J.*, 66, pp. 394–401.

83. Saxton, M. J. (1993), Lateral diffusion in an archipelago Dependence on tracer size, *Biophys. J.*, 64(4), pp. 1053–1062. doi: 10.1016/S0006-3495(93)81471-1.

84. Habasaki, J., Ngai, K. L. (2010), Molecular dynamics studies of ionically conducting glasses and ionic liquids: Wave number dependence of intermediate scattering function, *J. Chem. Phys.*, 133, p. 124505.

85. Tripol'skii, A. I., Strizhak, P. E. (2007), The effect of fractal dimension of porous catalysts on the activation energy of co oxidation, *Theor. Exp. Chem.*, 43, pp. 50–53.

86. Weeks, E. R., Crocker, J. C., Levitt, A. C., Schofield, A., Weitz, D. A. (2000), Three-dimensional direct imaging of structural relaxation near the colloidal glass transition, *Science,* 287(5453), pp. 627–631. DOI: 10.1126/science.287.5453.627.

Chapter 6

# Molecular Dynamics Simulations of Nanoporous Systems: Mechanism of Enhanced Dynamics of Ions

## 6.1 Introduction of Studies for Porous Systems

Porous materials have important roles in the recent development of nanotechnology. The possibility of the application of porous materials is widely spreading [1–5] because of their unique characteristics such as lightness and large surface area. In the research of nanoscience fields as well as in industry, porous materials are used as supports for catalysts [1]. They are also expected to be used for drug deliveries [2], where the solubility of the drug can be changed by porous adsorbents. Furthermore, porous materials are good candidates for electric devices, such as electrodes in solid-state batteries [3]. They are also expected to be useful for the storage of Li ions [4] and/or other materials. Porous structures also can be used as sheaves for the separation of materials [5]. Not only in applications, but porous materials can also be a target to examine the physical and chemical properties of materials with controlling or modifying them [6, 7], in fundamental research fields.

*Molecular Dynamics of Nanostructures and Nanoionics: Simulations in Complex Systems*
Junko Habasaki

ISBN 978-981-4800-77-8 (Hardcover), 978-1-003-04490-1 (eBook)
www.jennystanford.com

At the beginning of the study of porous materials, the silica system was a target to be examined by molecular dynamics simulations. Studies of silica and related alkali silicate are useful for understanding behaviors of the porous materials and give fundamental understandings for using them as devices and/or storages. In this chapter, some methods used to study porous silica in molecular dynamics simulations are introduced and then results of such modeling used for silicate systems are explained.

In nanoporous materials and their composites, enhancement of ionic conductivity has been reported as shown in Chapter 3 and possible mechanisms having different origins have been proposed so far. Especially, it is noteworthy that lithium silicate glass grown within pores and in composites also shows the enhancement as shown in Chapter 3.

In our previous works [6, 7], effects of introducing pores were examined by focusing enhanced ionic motion of Li ions in porous lithium silicates examined by molecular dynamics (MD) simulation in the constant volume and energy (*NVE*) conditions. Interestingly, enhanced dynamics having comparable characteristics to the composites shown in Chapter 3 were predicted. Therefore, the focus of the chapter is on clarifying the mechanism of enhanced diffusion found by MD simulations, although the mechanism found here is not necessarily the same as those in composites.

Usually, the high performance of porous materials tends to be attributed to their extremely high surface areas or structures of boundaries combined with the formation of pores [8]. However, there are many possibilities to be removed to decide what mechanism is important.

Since the introducing pore usually results in the structural changes of materials in the nanoscale and hence details of the structures and roles of pores need to be clarified. In this situation, MD simulations can play roles to clarify the mechanisms and structures in atomistic levels. It will be useful to show how possible mechanisms of the enhancement of the dynamics can be distinguished from each other in MD simulations in the case of the porous lithium disilicate for further examination of related systems.

In applications, the diameter of channels (pores) is an important factor to control the properties of porous materials. Porous materials can be subdivided into three categories by sizes of pores, set out by IUPAC as follows [9] as already mentioned in Chapter 2.

In the report, pores with widths exceeding about 50 nm are called macropores; pores of widths between 2 and 50 nm are called mesopores, and pores with widths not exceeding about 2 nm are called micropores. In Chapters 6 and 7, porous systems treated are mainly mesoporous and microporous ones.

Mesoporous silicate is experimentally prepared in several works. In the review by Alothman [10], the utility of silicate mesoporous materials for several purposes has been discussed. The purposes include supports for catalysis, separation, selective adsorption, and storage of gas or other guest molecules. It was argued that in silicate materials, well-defined pore sizes of about 2.0–10.0 nm and their range are better than that of microporous zeolites (<2.0 nm) for such applications. Systems are expected to take roles in environmental pollution control, especially for the uptake of heavy metals. Porous systems show a variety of morphology, which might affect the macroscopic behavior of the system. The properties of porous materials are affected by many factors besides the morphology.

## 6.2 Modeling of Nanoporous Materials

Characteristics of pores introduced in the systems depend on the methods of preparation. It is pointed out that channel-like structures are obtained by sol-gel methods [11], while more isolated pores can be obtained by chemical vapor deposition [12, 13], experimentally. In [1], three methods are compared to modify the size and/or shapes of pores. In the gel obtained from colloidal silica (see Chapters 8 and 9), we have examined a structure of gel, where networks of $SiO_4$ units are three-dimensionally connected infinitely by periodic boundary condition in MD simulations. In such cases, not only structures of silica networks, but also the structures and dynamics of the water trapped in the pore can be examined in parallel. By such analyses, the existence of open-pore structures is clarified. As found in this example, MD simulations are useful techniques to examine structures in porous materials as well as dynamics within them.

First, several methods for preparing porous models [14–18] used in the studies of porous silica will be summarized. After that, results of our MD simulations will be explained.

### 6.2.1 Modeling of Porous Silica

Kieffer and Angell [14] prepared porous silica by the expansion of the dense silica system, where both lengths of basic cell and positions of atoms are scaled. The method is performed under the *NVE* (constant number of particles, constant volume, and constant energy) conditions and was later refined by Nakano et al. [15] using *NPT* (constant number of particles, constant pressure, and constant temperature) condition to allow relaxing of the system volume during the run. The difference of porous structures in silicate in different extended conditions of MD runs will be discussed later.

A different method, a charge scaling procedure, was used for the preparation of a nanoporous silica model [16]. As in experiments, porous systems can be prepared by the aggregation process [17] and by a spontaneous gelation process [18]. A comparison of structures and mechanical properties obtained by several methods is found in ref. [19].

In ref. [18], starting from the nano-colloidal silica-water-salt system, spontaneous formation of the gel has been reproduced using fully atomistic MD simulations. The spontaneous formation of gel seems to provide a way to prepare the natural structures of the porous systems. The resultant porous structure of the silica part in the silica gel is shown in Fig. 6.1. In this figure, the formation of 3-D channels is clearly found from the structure without solvent. Besides, the structures of a solvent part filled in the channel can be used to learn the structure of pores and/or channels (see Chapters 8 and 9 for the details of the structure of silica and water parts).

The structure of the silica part can be used for the further analysis of porous systems. The porous structures after eliminating water and salt are found to be stable at least during several ns runs in both *NVE* and *NPT* conditions.

Another possible modeling method of porous materials in MD/MC may be to use a template of structures. The method using a random or systematic removal of atoms or units from the system can also be possible ways, although the modification of conditions is required to continue stable runs of the system.

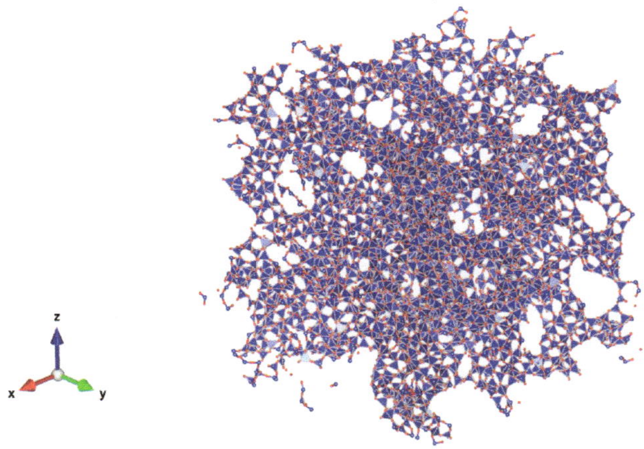


**Figure 6.1** An example of nanoporous silica taken from the gel spontaneously formed from nanocolloidal-silica—water-NaCl system obtained by MD simulation (see Chapters 8 and 9). Structures of water and NaCl are omitted in the present figure. Red and blue particles are for oxygen and silicon, respectively. Each $SiO_4$ unit is shown by a tetrahedron.

## 6.2.2 Modeling of Porous Silicate

Introducing pore is often accompanied with the changes of structures from ordered ones to disordered ones. Dynamics of the system will be affected by such structural changes and the effect is quite large. Actually, diffusion coefficients of ions in the ionically conducting materials in the crystalline state are known to be several orders slower than that in the corresponding glasses having the same composition [20–22] and this remarkable difference between glass and crystal has been confirmed by MD simulations of the lithium silicate systems [23], which are typical ionically conducting systems.

In the experimental situation of systems with nanocrystals or nanoparticles, the contribution of disorder introduced in the system with mixing other components will affect dynamics considerably. Therefore, before discussing the effects caused by the modification of the system by pores, elimination of the effects in different ordering is necessary.

For this purpose, we prepared the porous system with starting from the system in the glassy state [6, 7], so that the effects of introducing disorder is removed *a priori*. That is, porous lithium disilicate systems were prepared by the expansion of the glass, which was obtained by the rapid cooling of the corresponding melt. By this procedure, one can eliminate effects of different contents of the crystalline phase and glassy phase, which might affect the enhancement. The functional form and description of the potential models used in the MD simulations are shown in the Appendix.

In Chapters 6 and 7, density $\rho$ is given in g $cm^{-3}$ and time is in ps, unless otherwise stated. We may use non-SI units, case by case, for several reasons. One of the reasons is the convenience to treat problems in intuitively easier ways. For example, lengths are given in Å (1 Å = 0.1 nm), because it is convenient to treat atomistic sizes. Porosity (defined by the ratio of the volume of space to the total volume, that is the fraction of void), $P_0$ (in %) was obtained from MD simulations (after equilibration at each density) and is represented as the function of each density, $\rho$ for porous systems, including the system before expansion.

## 6.3 Enhanced Dynamics in Lithium Disilicate Systems in NVE Conditions

As discussed in Chapter 2, enhancement of the dynamics (or an opposite case) has been reported for several porous systems and/or related composites as shown, for example, in [24].

For a better understanding of the mechanism of the enhancement of the dynamics in a nanoporous system, molecular dynamics (MD) simulation of porous lithium disilicate has been performed in refs. [6, 7] and is the focus of the present chapter. As shown in detail in the next section, enhancement of diffusion of Li ions is found in the porous lithium disilicate examined in *NVE* conditions.

### 6.3.1 Possible Origins of Enhancements

To explain the mechanism of the enhancement in porous ionic conductor, including several composites such as LiI in alumina or silica, several different scenarios have been proposed.

Although the following concepts and/or explanations discussed for nanocomposites (Chapter 3) are not necessarily directly applicable for porous lithium disilicate treated in *NVE* conditions, it is worth to examine because the existence of some common mechanisms underlying can be expected [8, 25].

So far, enhanced dynamics in several porous systems have been explained by

1. a concept of charge transfer and space charge effects [26, 27];
2. an effect of epitaxial strain in opening diffusion paths [28];
3. the influence of atomic and electronic reconstruction at the interface in stabilizing new phases with increased carrier concentration [29, 30];
4. the percolation theory with assuming high conductivity regions on the boundary [31].

Some of the possible origins (like impurity such as water) explained in Chapter 3 can be excluded in well-controlled conditions in MD.

Most of these explanations seem to be related to the characters of the boundaries of the particles of insulator rather than related the porous nature of the components of composites or mixtures. However, a question remains for the role of nanoparticles and/or nanoporous materials.

## 6.4 MD Simulations of Porous Lithium Disilicate

### 6.4.1 Methods of Simulations for Porous Lithium Disilicate Systems

First, the characteristics of the potential parameters used in MD simulations will be discussed, because the validity of the mechanism discussed in MD simulation depends on the quality of them considerably.

Our potential parameters [32] have been previously developed based on the *ab initio* MO calculations and the model covers a wide range of potential surfaces. Based on our potential model, lithium disilicate system and related systems have been examined successfully from many points of view by several groups including our groups [8, 12, 32–39] in the molten, glassy, crystalline states. Therefore, it can be used for the comparison of systems with different states, including porous ones although it was not optimized for the study of porous systems. Of course, an examination of applicability is desirable for each purpose.

For the preparation of lithium disilicate systems with several different densities (porosities), the system and the coordinate were expanded gradually. That is, an expansion of the system by scaling of factor 1.005 times (for both the axis length and the coordinate) was repeated until the density to be examined was obtained. Densities $\rho$ examined are 2.47 (original), 2.30, 2.13, 1.98, 1.84, and 1.58 (density is given in g $cm^{-3}$). The experimental density of the system in the glassy state was reported to be 2.35 [40]. After the expansion of the system, each system was equilibrated during 300,000 steps run in the *NVT* condition. Then further 300,000 steps run were done in the *NVE* condition. If temperature increased during the *NVE* run (due to non-equilibrium nature of the system), short *NVT* runs (~30,000 steps using Gaussian thermostat) were inserted between *NVE* runs. The run for 300,000–4,000,000 steps, which was performed in the *NVE* condition after the equilibration, was used for further analysis. For temperature dependence of the dynamics (600–1000 K) in porous materials, simulations have been started from the porous systems at 600 K and the same system volume was used at each target temperature. In some cases, additional runs for different density and/or at different temperatures have been performed.

### 6.4.2 Relation between Porosity and Density in Porous Lithium Disilicate Systems

In refs. [6, 7], the values of the porosity $P_0$ were determined by geometric considerations using the ionic radius. Here we used

Si: 0.40, O: 1.21, and Li: 0.8 as atomic or ionic radii (in Å), which are consistent to the maximum position of the first peak of $g(r)$ for Si–O (1.63 in the original system at 600 K) and Li–O (2.04 in the original system at 600 K) pairs. The method used is not for "available space" but the "precise amount of free space." The method is not necessarily suitable when the overlaps of atoms are non-negligible but seems to be suitable to treat porous systems.

For the systems with $\rho$ = 2.47, 2.13, 1.98, 1.84, and 1.58 (in g $cm^{-3}$) at 600 K, porosities are determined to be, 56.62, 62.59, 65.34, 67.80, and 72.29, respectively. The porosity, $P_0$ (in %), was found to be (negatively) proportional to the density $\rho$ (in g $cm^{-3}$) of the system, that is,

$$P_0 = -17.57\rho + 100.08. \qquad (6.1)$$

Hence, the results related to the different porosities of the systems will be shown simply as the function of the density.

### 6.4.3 Trajectories of Ions and Atoms in a Porous Systems Compared with the Original System

In Figs. 6.2a,b, trajectories (Li ions, Si and O atoms) in original and porous lithium disilicates during 100 ps run at 600 K are shown for a slice with a width of 10 Å of MD cell, respectively. The density of the porous system is $\rho$ = 1.98, where the maximum of the diffusion coefficient at 600 K is found as shown later.

In Fig. 6.2b, enhanced dynamics are observed in the bulk part and not necessarily remarkable at the boundary of the clear pores. They are accompanied with the larger motions of Si and O atoms than the original system. In Fig. 6.2c, trajectories for longer runs (5 ns) in the original system are shown, where the MSD of Li ions is comparable to the value found in (b). In Figs. 6.2b,c, ion channels are observed by the overlap of trajectories of Li ions with different time scale region. Thus, the enhanced diffusivity in the porous system is found. Some ions are located for a long time, while fast ions cooperatively move in the channels. Therefore, the enhanced ionic motion is heterogeneous one as known for the original system as well as other ionic systems.

Details of the heterogeneous dynamics in porous lithium disilicate systems will be discussed in Chapter 7.

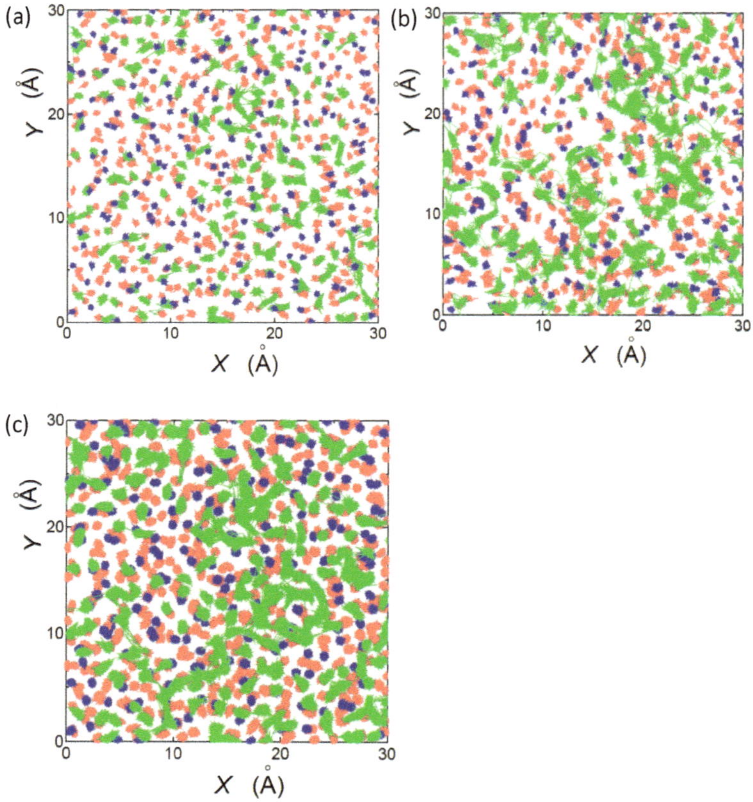


**Figure 6.2** (a) Trajectories of ions and atoms during 100 ps (100,000 steps) run of original $Li_2Si_2O_5$ glass at 600 K for a slice with a thickness of 10 Å of MD cell. (b) Enhanced dynamics found in the trajectories of ions in porous lithium disilicate during 100 ps run near the maximum of the mobility at 600 K with $\rho$ = 1.98. Green: Li ions, Blue: Si atoms and Red: Oxygen atoms. Not only motion of Li ions, mobility of O and Si atoms in the porous system are larger than the original system. (c) Trajectories of ions and atoms during 5 ns run of original $Li_2Si_2O_5$ glass at 600 K for a slice with a thickness of 10 Å of MD cell of the system, where the MSD (10 Å$^2$) of Li ions is comparable to the Li ions in (b). In each case, data are plotted with interval of each 500 fs. Enhanced dynamics are clearly found in porous lithium disilicates. In these figures, diffusive paths are not necessarily located near the boundaries. Trajectories of Si and O are spread in the porous system and it suggests the loosening of cage formed by O atoms around $Li^+$ ions.

### 6.4.4 Enhanced Dynamics and Existence of Maximum in NVE Conditions

As shown in the following sections, dynamics of ions in porous lithium silicate in *NVE* conditions show common characteristics as those found in other porous ionic systems treated in Chapter 3. It includes the existence of the maximum of the enhanced diffusivity in the medium density (or composition) region and smaller slopes in Arrhenius plots of the diffusion coefficients. Since the enhanced dynamics are observed in several composites of lithium silicate and ZnO [41] as well as in the system LiI-$Al_2O_3$ [42], it can be expected common physics are underlying these behaviors [8, 25]. Common behaviors do not necessarily guarantee that the same explanation holds in all systems. However, the methodology of separating the possible mechanisms and examining the nanostructures and nanoionics in this system will be useful in other composites. In the field of nanocharacterization, where molecular dynamics simulations can be expected to contribute, it seems that new researchers are waited to enter.

## 6.5 Comparison of MSD of Li Ions in the Porous and Original Systems

From the changes in the trajectories, it was suggested that the caging dynamics are affected by the decreased density (and porosity). In this case, we can expect that the MSD of the Li ions, as well as those of Si and O atoms, will change. To separate the possible origins and quantitative assessment of the enhancement of diffusivity (or conductivity) with porosities, time developments of MSD of systems with different density were examined.

### 6.5.1 Several Characteristic Regions of MSD of Li Ions in Original Lithium Disilicate Glass

Mean-squared displacement (MSD) of ions or atoms can be obtained from the MD simulations data using the following relations.

$$< r^2(t) > = \left\langle \{\sum_{i=1}^{N}(\mathbf{r}(t)-\mathbf{r}_i(0))^2\}/N \right\rangle \qquad (6.2)$$

where $\mathbf{r}_i$ is the positional vector

As discussed in Chapter 5, the characterization of the motions of Li ions in each time region has been done with the help of van Hove functions, fractal dimension of random walks, intermediate scattering functions for both lithium metasilicate and lithium disilicate [6, 8, 33, 39, 43]. The result is briefly summarized below because these time regions are compared with each other in porous systems.

Four distinct time regions for MSD of Li ions are found in the original (before expansion) $Li_2Si_2O_5$ system in the glassy state. MSD of Li ions at 600 and 800 K in this system is shown in Fig. 6.3.

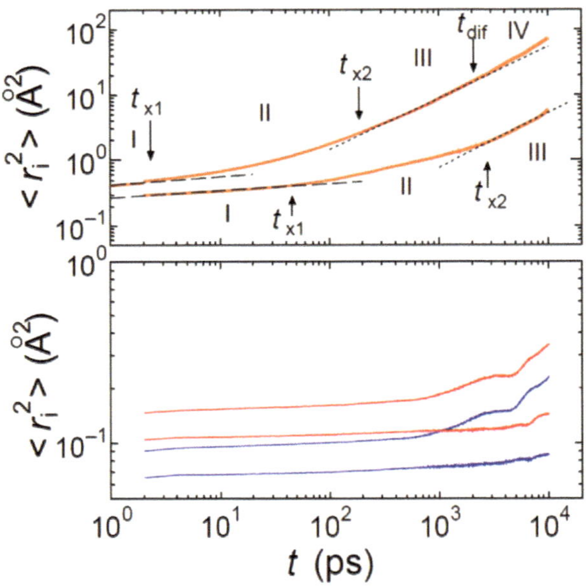


**Figure 6.3** Top: MSD of Li ions in original Lithium disilicate glass at 600 K (lower curve) and 800 K (upper curve). Dashed and dotted lines (fitted to power laws) for Li ions at each temperature are for NCL and power law regions, respectively. Bottom: MSD of Si (Blue curves) and O (Red curves) at both temperatures. At 600 K, nearly flat (NCL) region is expanded to ~30 ps and the power law region starts at ~3 ns. Flat region of MSD for Si and O continues up to ~ns at both temperatures.

**Region I (NCL)**: After the initial increase of MSD due to the local motion of the ion site, a nearly flat region assigned to the nearly constant loss (NCL) is found. This time region is located between approximately 0.2 and 2 ps in the case at 800 K where the MSD increases with time as $t^{\alpha}$ with $\alpha \approx 0.07$. In this time region, ions are still caged by surrounding oxygen atoms and further shells of ions and atoms. Region I is characterized as the caging region, where the anharmonic motion of ions with intermittency was observed [44, 45].

Changes in this characteristic time region are useful to understand the mechanism of the enhancement of dynamics of ions. If the time region becomes longer, this means that the cage is tight, while if the region becomes shorter, it means that the caging is loose and jump motions to the next shell begin. Therefore, the length of this time region is related to the jump rate of the caged ion.

**Region II**: Primitive jumps (the first jump of each ion after the starting observation for some ions) are found in this intermediate time regime of about $t_{x1} < t < t_{x2}$. The MSD rises more rapidly than $t^{\alpha}$ of the regime (I) in a double logarithmic plot. This increase of MSD is nearly proportional to time as easily found in the linear scale plot.

**Region III**: Power law behaviors of MSD are found in the region $t_{x2} < t < t_{\text{dif}}$. That is, MSD depends on $t^{\theta}$ with $\theta < 1$. Both fast and slow ions are found here and the slope $\theta$ is affected by the back-correlated motion of jumps considerably in the glass. That is, the region proportional to time in region II ends by overlaps of jumps returned to original sites and jumps to further shells. In some porous systems with large $\theta$ (>0.8), the linear function to time can be a good approximation for representing MSD.

Depending on the purpose, a correction might be required for the contribution of motions within a cage to MSD for examining the slope in the double logarithmic plot.

**Region IV**: Diffusive region of ions is found after $t_{\text{dif}}$, where MSD reaches the square of the typical neighboring distance of ion sites and the MSD shows a linear $t$ dependence in that region.

With decreasing temperature, each time region becomes longer and longer. For example, the diffusive regime is found after 2 ns at 800 K for Li ions, while it is found after ~15 ns at 600 K.

From the slope after $t_{\text{dif}}$, the diffusion coefficient of ions can be determined by the Einstein equation [46]:

$$D = \frac{1}{6}\lim_{t\to\infty}\frac{d}{dt} < [\mathbf{r}_i(t) - \mathbf{r}_i(0)]^2 >, \tag{6.3}$$

where $D$ is the diffusion coefficient and $\mathbf{r}_i$ is the positional vector.

## 6.5.2 Mean-Squared Displacement of Si and O Atoms in Original Lithium Disilicate Glass

Motions of Si and O of this system are shown in the lower panel of Fig. 6.3. For both 600 and 800 K, they are immobile for a long time and hence the cage is stable during this time period. A slight increase in the mobility for these species is found at ~1 ns at 800 K.

Diffusion coefficients are fundamental transport properties, which are closely related to other ones such as conductivity [47–50] as already discussed in Chapter 5.

### 6.5.2.1 Dynamical changes toward the maximum of the diffusion coefficient with the expansion of the system

The MSD of porous systems obtained in *NVE* condition shows a remarkable difference from the original glass. In Fig. 6.4, the time dependence of the MSD of Li ions for porous systems is compared with the original glassy system. In the porous systems, MSDs levels off earlier than that in the original glass and hence the NCL region is shortened. That is, the change in the dynamics of ions starts at the early time region for caged ion dynamics.

It is worth to mention that the motion affected by introducing pore is not only for Li ions but also for network-forming atoms as shown in the bottom part of the figure. In porous systems showing high-diffusivity of Li ions, levels of Si and O are found to be larger than the original system and no flat region of O atoms is observed after 1 ps. Motions of network-forming atoms are then non-negligible even in the short time region. Because these atoms are forming cages for Li ions, this result also means that the loosening of the cage is contributing to the enhancement of mobility of Li ions. Such changes in coordination polyhedra naturally increased in the jump rate. These features are consistent with the observation in trajectories shown in Fig. 6.2 and accompanied with the decrease in the coordination number $N_V$ (= $x$

in $LiO_x$ structure), as shown in Fig. 6.5a. In this figure, the position of the maximum of the peak is found at 5 before the expansion of the system and gradually decreases. This change accompanied with the increase of the diffusivity.

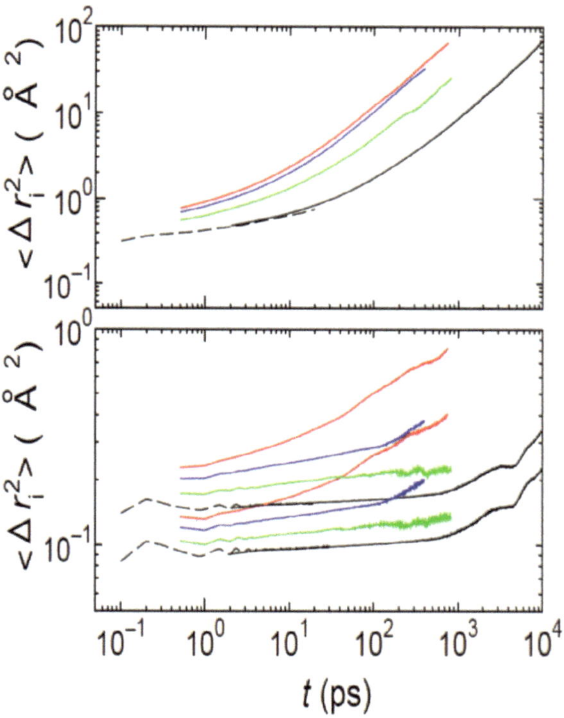


**Figure 6.4** Shortening of the caging (NCL) region in porous systems. Top: MSD of ions and atoms in lithium disilicate in porous systems at 800 K (upper curves). Upper panel: Li ions for $\rho$ = 2.47 (black) (original), 2.30 (green), 2.13 (blue) and 1.98 (red) from lower to upper. The diffusivity of the system increases in this order. In this figure, long-time data for the $\rho$ = 2.47, which were obtained by a larger time step (4 fs), are smoothly connected to those for short time scales (dashed curves) obtained by a smaller time step (1 fs). Bottom: MSD of Si (lower four curves in the beginning part) and O (upper four curves in the beginning part) of the same systems. The same color as that for MSD of Li ions is used for each atom. Reprinted from Habasaki, J. (2016), Molecular dynamics study of nano-porous materials—Enhancement of mobility of Li ions in lithium disilicate in *NVE* conditions, *J. Chem. Phys.*, **145**, 204503(1–11), with the permission of AIP Publishing.

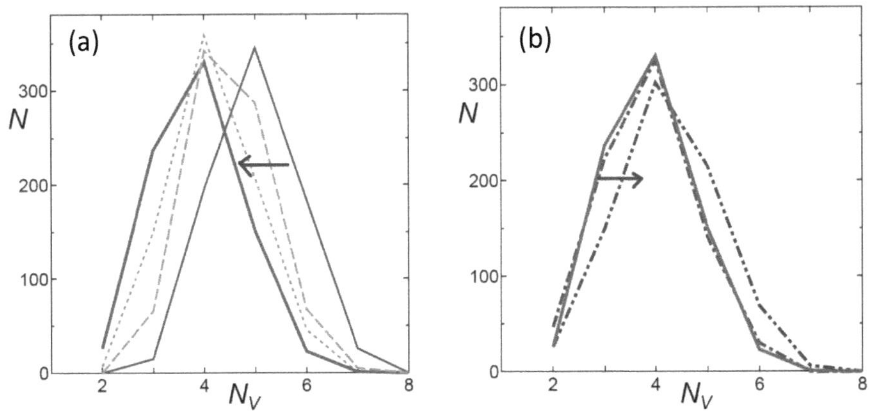


**Figure 6.5** Topological change of the coordination polyhedra (O atoms around Li ion) by introducing pores with decreasing density. (a) Changes of the distribution of the coordination number ($N_V$) at 800 K. Green (solid) line: for the original lithium disilicate glass. Pale green (dashed) line: for the porous system with $\rho$ = 2.30, orange (dotted) line: for $\rho$ = 2.13 and red (thick solid) line: for $\rho$ = 1.98. An arrow (blue) in this figure means the trend with decreasing density. (b) Further changes in the coordination number. The red one is the same one as in (a). The distribution connected by dotted dashed lines is for $\rho$ = 1, and that connected by two dotted dashed lines is for $\rho$ = 1.58. The trend shown by the arrow is just opposite to that found in (a). That is, the loosening of the cage (by decreasing overlaps of polyhedra occurs in (a), while the tightening of it occurs in (b). Reprinted from Habasaki, J. (2016), Molecular dynamics study of nano-porous materials—Enhancement of mobility of Li ions in lithium disilicate in *NVE* conditions, *J. Chem. Phys.*, 145, 204503(1–11), with the permission of AIP Publishing.

#### 6.5.2.2 Further decrease of the dynamics with the expansion of the system

With the further decrease of the density ($\rho$ = 1.84 and 1.58), the coordination number of $LiO_x$ structure tends to increase (Fig. 6.5b).

With this structural change, the decrease of the MSD is found (see Fig. 6.6). That is, MSDs of Li ions, Si and O atoms decrease and the NCL region of these species becomes longer again. This slowing down also starts from the early time region, implying the tightening of the cage. This structural change is accompanied with the gradual structural changes with a formation of larger voids as shown later.

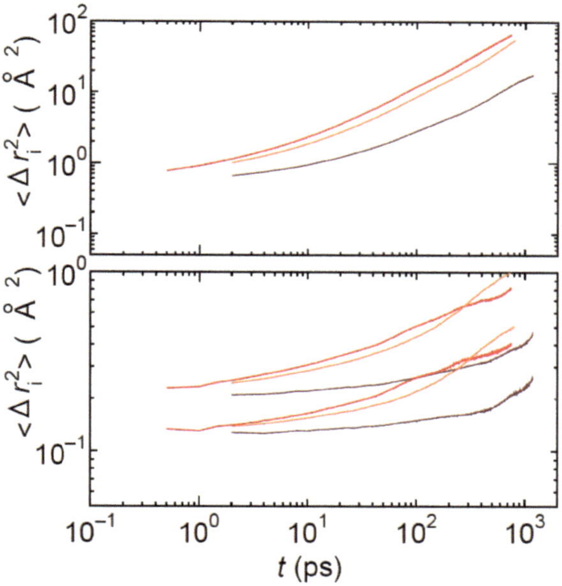


**Figure 6.6** Dependence of MSD on density (porosity) of $Li_2Si_2O_5$ systems at 800 K. Upper panel: Li ions for $\rho$ = 1.98, 1.84 and 1.58 from upper to lower. The diffusivity of the system decreases in this order. Lower panel: Si (lower three curves) and O atoms (upper three curves) Density is 1.98, 1.84, and 1.58 from upper to lower for each species. Reprinted from Habasaki, J. (2016), Molecular dynamics study of nano-porous materials—Enhancement of mobility of Li ions in lithium disilicate in NVE conditions, *J. Chem. Phys.*, 145, 204503 (1–11), with the permission of AIP Publishing.

## 6.6 Common Behaviors of Porous Lithium Disilicate and Composites Showing Enhanced Dynamics

The next step is to treat the transport coefficients quantitatively. For this purpose, diffusion coefficients of ions and atoms are plotted against the density of the system (Fig. 6.7).

In Fig. 6.7a, diffusion coefficients of Li ions determined by the Einstein equation are plotted against the density of the systems at 600, 700, and 800 K. The maximum of the diffusion coefficient is predicted at around $\rho$ = 1.98, where the diffusion coefficient is more than one order higher than the original system at 600 K.

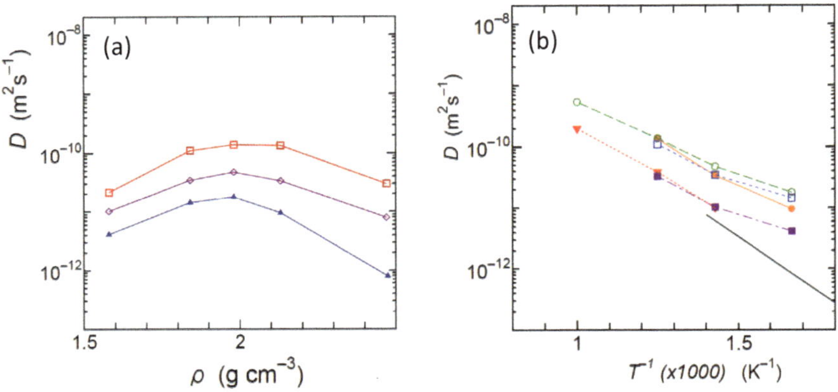


**Figure 6.7** (a) Diffusion coefficients of Li ions in porous lithium disilicate in the *NVE* condition. The largest density (the right end point) corresponds to the original system before the expansion. Blue triangles: 600 K, purple diamonds: 700 K and red open squares: 800 K. The maxima are found at around $\rho$ = 1.98 (in g $cm^{-3}$) for all temperatures, where the diffusion coefficients are predicted to be one order higher than the original systems. The enhancement is larger at lower temperatures in the log-scale plot. (b) Temperature dependence of diffusion coefficients of Li ions for each $\rho$ value. Inverse triangles (red): $\rho$ = 2.47, system in the glassy state without expansion. Filled circles (orange): $\rho$ = 2.13, open circles (green): $\rho$ = 1.98, open squares (blue): $\rho$ = 1.84 and filled squares (purple): $\rho$ = 1.58. The solid black line indicates the approximate position of experimental values for lithium disilicate glass gathered in Ref. [49]. Smaller negative slopes are found as in other porous composites. Reprinted from Habasaki, J. (2016), Molecular dynamics study of nano-porous materials—Enhancement of mobility of Li ions in lithium disilicate in NVE conditions, *J. Chem. Phys.*, 145, 204503(1–11), with the permission of AIP Publishing.

As already shown qualitatively by MSD, the maximum of the transport properties was found for the diffusion coefficient of Li ions in porous lithium disilicate systems simulated in the *NVE* condition.

The shape of curves connecting different densities (porosities) slightly changes with temperature. This trend is characterized mainly by the larger temperature dependence of the diffusivity in the large density region. The existence of the maximum is usually found in the composition dependence of the composites showing enhanced dynamics as shown in Chapter 2 and/or in [41, 42].

Temperature dependence of the diffusion coefficients in these systems is shown in Fig. 6.7b. The similarity with other porous composites is also found in the temperature-dependent behavior. The result of the original system shown by the dotted line is comparable to the trend shown by the solid one, which represents the approximate positions of experimental values of Li ions in the original glass [51]. Interestingly, smaller negative slopes in the temperature dependence in porous systems are found in this figure. As in porous composites showing enhanced diffusions and/or conductivities, the $D$–$(1/T)$ plot shows the smaller negative slope than the original system. For example, see Ref. [52] for common trends for several systems. This fact is promising to use the porous system as high conductivity materials in the lower temperature region.

These common trends also suggest the possibility of the existence of the common physics for the origin of the enhancement and these characteristics. That is, we can expect that porous materials act as supports of mobile ions in composites and they will cause a modification of caging dynamics, although the mechanism should not be the same. A porous system will also modify jump paths as shown later. Of course, the difference in the conditions used in MD simulations and experimental ones also should be considered, since experimental results tend to be obtained in the *NPT* condition instead of the fixed volume in the *NVE* condition in the present system for further comparison.

Comparable behaviors of the conductivity to the diffusion coefficient for the existence of the maxima and/or similar trend in the slopes in Arrhenius plots are naturally expected.

### 6.6.1 Contribution of Other Factors: Percolation Theory, Fractal Dimension of Random Walks and Cooperative Jumps

How can we distinguish the possible mechanisms, which affect the MSD and hence the diffusion coefficient? Here we will examine the characteristic time scales defined for MSD to clarify the origin of the enhancement and the cause of the maximum in the porous lithium disilicate. Causes of the asymmetry of the curve for the diffusion coefficient against density plot (Fig. 6.7a) are clearer at lower temperatures, and dynamics are more enhanced in a longer time scale in some cases. These findings suggest the existence of another factor (or factors) playing roles to determine diffusivity. One of the possible explanations concerns the geometrical changes of the trajectories, related to the structure of the jump paths and cooperative jump motions.

This is because the exponent $\theta$ is a function of the fractal dimension of the random walk [53], $d_w$, which can be used as an index to represent the complexity of trajectories [53, 54]. Here, the exponent contains only the geometrical term. For a free random walk, one can expect $d_w = 2$, while for more localized motions the value becomes larger. In the trajectory of fast ions, the trajectory becomes more linear, and the value becomes less than 2 [54]. Larger $d_w$ value means that the trajectory is more complex and the ion tends to be localized and so that the small value of $\theta$ (larger value of $d_w$) results in the retardation of the time scale of $t_{dif}$.

### 6.6.2 Fractal Dimension of Random Walks

The fractal dimension of the random walk, $d_w$, was determined from the trajectories of each ion by using a divider method [8]. This dimension represents the complexity of the trajectory and is defined by the following equation:

$$N_L = A' L_r^{-d_w} \tag{6.4}$$

In this analysis, trajectories were obtained by the MD simulation during a run of time $t_{run}$. For the determination of $d_w$ for heterogeneous dynamics, the long run should be used for sampling of rare events. If the time scale is long enough, exchanges between fast and slow ions will also be observed.

### 6.6.3 Multifractal Random Walks Found in Lithium Silicates

In the lithium metasilicate, the slope of the $N_L$–$L_r$ plot was found to change at around 3 Å at lower temperatures. The $d_w$ was named $d_{w1}$ for a shorter length scale (<3 Å) and $d_{w2}$ for a longer length scale ($3 \leq L_r < 10$ (Å)). From the theories of random walks, $\theta = 2/d_\mathrm{w}$ is expected for a simple random walk as already mentioned [54]. However, the existence of different length scales found here means that the value is for the average of slow and fast ions forming multifractal random walks [55–59]. Since the fractal dimensions of the walks are closely related to the structures of jump paths, it is interesting to consider the relation with percolation theories [60–62].

For the porous media, the effects of the enhancement of transport by the contribution of low-frequency elastic waves also have been pointed out in [63].

### 6.6.4 Percolative Aspect of Jump Paths

Generally, the concept of percolation is quite useful. However, in the case of the enhancement of porous lithium disilicate system examined in the *NVE* condition, the main cause of the enhancement, as well as the existence of the maximum of the conductivity, cannot be explained by this kind of theory alone, although the effect concerning the path still exists. Instead, it can be mainly explained by the changes in the caging dynamics at least in the present system in the *NVE* condition. Since the slow dynamics in the densely packed systems are affected by the back correlated motions of ions, the geometrical characteristics of paths are the non-negligible factors to control dynamics, as well. Here we consider the percolative aspect of the jump path of lithium ions in the porous lithium disilicate system. To learn the effect of geometrical factors such as correlation among jump motions of ions with separating from the changes in caged ion dynamics, characteristic times of MSD are examined. Effects of caging dynamics are shown by the changes in $t_{x1}$, while the changes in the power law exponent $\theta$ of MSD is shown by the time scale separation between $t_{x2}$ and $t_\mathrm{dif}$, based on the physical meaning of each stage of ionic motion (see Chapter 5). In the following section, we will make a quantitative assessment of these factors and relations between them.

### 6.6.5 Characteristic Times in Porous Systems

To understand how changes in dynamics are brought, to specify time scales of motions in porous lithium disilicate is useful and they can be quantitatively examined by characteristic times $t_{x1}$, $t_{x2}$ and $t_{dif}$ (in ps), which separate different regions found in MSD for Li ions.

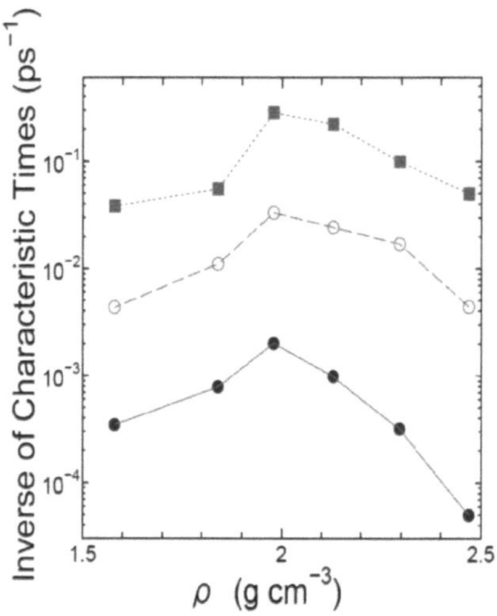


**Figure 6.8** Inverse of the characteristic times (in ps$^{-1}$) found in MSD of Li ions plotted against density of the system for lithium disilicate at 600 K. Filled squares, green: $1/t_{x1}$. Circles, blue: $1/t_{x2}$. Filled circles, dark blue: $1/t_{dif}$. Maxima are found even in the shortest time scale. Each error of datum in this figure is estimated within the size of marks, except for the value at the lowest temperature for $1/t_{dif}$, where the value is for the largest limit, due to the overlap of an aging process. Thus, the shortening of the time scales contributes to the enhancement of the diffusive motion. See also Chapter 7 for further explanation of this figure. Reprinted with permission from Habasaki, J., Ngai, K. L. (2018), Heterogeneous dynamics in nanoporous materials examined by molecular dynamics simulations—effects of modification of caged ion dynamics, *J. Non-Cryst. Solid*, 498, pp. 364–371. Copyright (2018), Elsevier.

As shown in Fig. 6.8, the inverse of the characteristic times (in ps$^{-1}$) of Li ions in the porous lithium disilicate system obtained by the *NVE* condition shows the maximum at $\rho$ = 1.98 (g cm$^{-3}$) at 600 K. The shortening and lengthening of the characteristic time $t_{x1}$ found to correspond to the loosening and tightening of cages, respectively. For the latter behavior, more details will be shown later. Here our attention is turned toward the maxima. Changes in dynamics are more enhanced on a longer time scale. Especially, the change of $1/t_{dif}$ between $\rho$ = 2.15 and original system is larger compared with those for $1/t_{x2}$ and $1/t_{x1}$ at high-density region. Asymmetry of the curve similar to this situation was clearly found in the diffusion coefficient plotted against density as shown in Fig. 6.8. This result is consistent with the observation that the activation energy of the $t_{\rm dif}$ is comparable to that of the diffusive motions [45].

Similar behaviors with maxima were observed at 700 and 800 K as shown in Fig. 6.9, The maxima were found even for $1/t_{x1}$ at higher temperatures, although the changes in short time scale at 800 K are small.

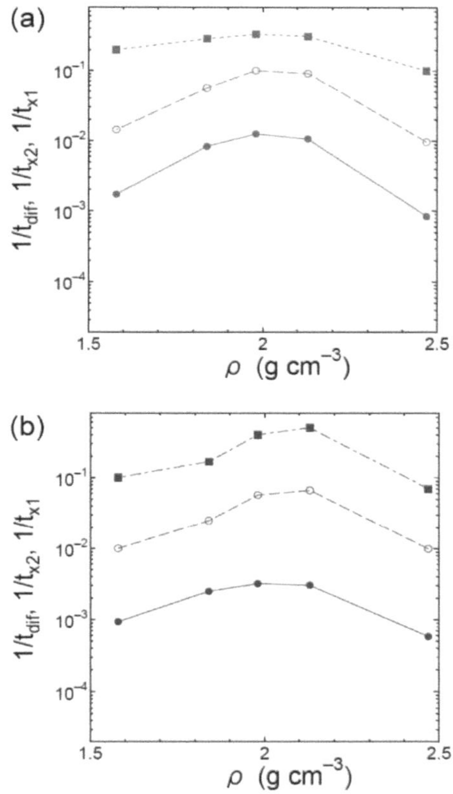


**Figure 6.9** Inverse of the characteristic times (in ps$^{-1}$) found in MSD of Li ions plotted against density of the system for lithium disilicate at (a) 800 K and (b) 700 K. Upper, medium and bottom are for $1/t_{x1}$, $1/t_{x2}$ and $1/t_{\rm dif}$, respectively. The maximum of the values has been already found in the shortest time scale even at higher temperatures. The change near the dense part seems to be weakened at the higher temperature.

## 6.7 Relationships among Multifractal Random Walks, Heterogeneous Dynamics, and Power Law Exponent $\theta$, of MSD

Although the heterogeneous dynamics are considered as the multifractal random walks [55–58] having different length scale (and time scale) regions [8], the relation mentioned in 6.6.3 between $\theta$ and $d_w$ is expected to hold for mean behaviors of heterogeneous dynamics, although large fluctuations are found frequently due to coexistence of fast and slow motions of particles [33, 64].

The concept of the multifractal random walk is useful to understand the complex behaviors, not only for ionics but also for other time series in many systems found in geophysics, biology and economics, *etc.* Therefore, it seems to be useful to understand the possible origins of such behaviors. Recently one molecular motion in several systems [65, 66] is examined and they are interesting examples to be examined concerning the multifractality. In ionic systems, the complex motions of ions, where exchanges among fast and slow dynamics construct the multifractal walks [8, 70–72], have been attributed to the convolution of truncated Lévy distribution of displacements [67–69] and geometrically correlated jumps.

### 6.7.1 Percolative and Cooperative Aspect of Dynamics

In Fig. 6.8, an asymmetry of the plot is found in a longer time scale behavior and larger change is found in the dense region marked by line with arrows. This asymmetry found in the dynamics in porous systems cannot be explained by the changes in the caging dynamics alone. Instead, the large change in $1/t_{\text{dif}}$ was attributed to the geometrical changes of the trajectories, related to the structure of the jump paths and correlation among successive jump motions. This conclusion was drawn from the following characteristics of $t_{\text{dif}}$ found in MSD.

As discussed in Chapter 5 and in [8], in the power law region of MSD, both fast and slow ions are usually found in the original glass and the slope $\theta$ is affected by the forward or back-correlated motion of jump motions of ions. For the mean behavior of fast and slow ions, $\theta = 2/d_w$ is expected, where the exponent $\theta$ is a function of the fractal dimension of the random walk [53], $d_w$. The value can be used as an index to represent the complexity of trajectories [8]. Larger $d_w$ value means that the trajectory is more complex one and ion tends to be localized and so that the small value of $\theta$ (larger value of $d_w$) resulted in the delay of $t_{dif}$.

The result is useful to understand the dynamics in the glass, where the ions tend to be located near the bottom of the potential surfaces. For the dynamics of ions in the glassy lithium disilicate system, change in the characteristic times related to the slope of the MSD in the power law region contributes to the slowing down of the dynamics considerably and this situation is changed by introducing pores.

Changes in $\theta$ were determined from the power law region of MSD and plots against density will be shown in Chapter 7. As discussed there, separation of different contributions of caging and the power law exponent $\theta$ examined in the present work is consistent with the previous results [70] found in an ionic liquid, EMIM (1-ethyl-3-methyl imidazolium)-$NO_3$.

### 6.7.2 Structural Changes Related to Enhancement of Dynamics

Hereafter, we consider what kind of structural changes are found at the atomistic level by introducing pores. To consider the origin of fast ion dynamics, the concentration of Li ions is one of the important factors to be considered. This is because, for composition dependence of $M_2O$-$SiO_2$ (M=Li) glasses, it is known that the system shows higher diffusivity and corresponding conductivity with increasing contents of alkali metal ions. In the high-alkali-content system, the contribution of fast ions with cooperative jumps increases. In Fig. 6.10, a pair correlation function $g(r)$ of Li–Li pairs of the porous system ($\rho$ = 1.98) at 600 K is compared with that in the original system. The peak positions of the first and second peaks shift toward the right and peak heights become smaller by decreasing density. Usually, the formation of ion channels is observed in the region with a high density of ions in the original system. Therefore, one may expect that the formation of the high-density region of Li ions as a cause of the enhancement. However, this is not the case. Instead, enhancement is mainly caused by the loosening of the cage. Li ions in the porous system are more separated by decreasing density as shown in Fig. 6.10.

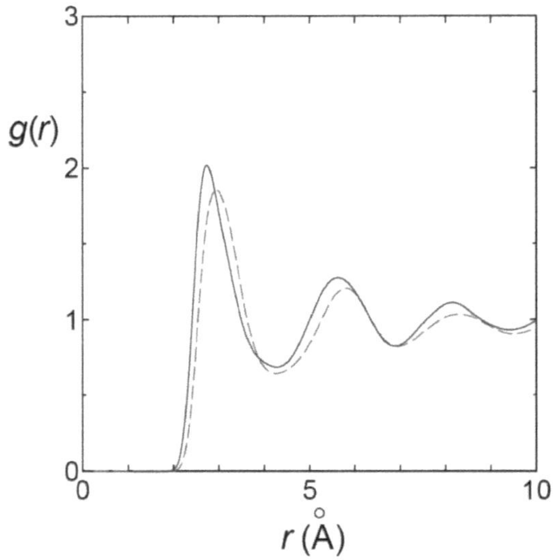


**Figure 6.10** Pair correlation function $g(r)$ for Li–Li pair in original $Li_2Si_2O_5$ glass (Blue, solid curve) and porous system (red dashed curve) ($\rho$ = 1.98) both at 600 K. In the latter, maximum diffusivity was observed.

## 6.8 The Role of the Caged Ion Dynamics in the Enhancement of Dynamics

### 6.8.1 The Concept of Geometrical Degree of Freedom

Structure of cage in lithium silicate system can be well characterized by the structure of $LiO_x$ [73] as well as structures of silica part using the concept of geometrical degree of freedom [74], which is schematically shown in Fig. 6.11.

The role of the cage in the ionic system is rather general. A quite similar discussion is applicable for ionic liquids. In our previous works [75] for ionic liquid, 1-ethyl-3-methyl imidazolium nitrate (EMIM-$NO_3$), we have shown that the mobility of ions is closely related to both the change of the geometrical degree of freedom of the coordination polyhedra and total number of bonds (including fictive ones for neighboring (contact) ions or atoms). The combination of these concepts can explain both characteristic temperatures, $T_B$ and $T_g$, of the system. A loss of the geometrical degrees of freedom occurs with the closed packing of coordination polyhedra in a shell, while the saturation of the number of the constraints (bonds, contact ion pairs) is found near $T_g$, which is related to the concept of the rigidity percolation of (fictive) bonds [74].

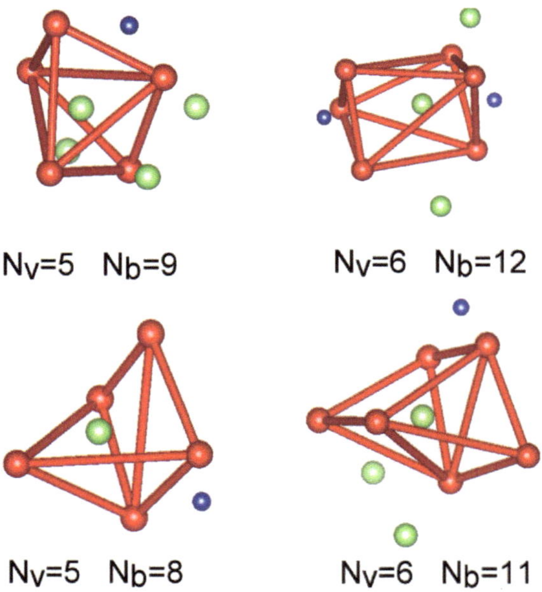


**Figure 6.11** Examples of $LiO_x$ polyhedra and their characteristics for the coordination number (number of vertices), $N_v$ = 5 (left) and 6 (right). Structures are taken from lithium disilicate systems obtained by MD simulations. Other Li ions and Si atoms within 2.8 Å from each central Li ion are also shown. The number of contact O pairs (fictive bonds), $N_b$, are counted. Geometrical degree of freedom [74] is lost when $(3N_v\text{-}6)\text{-}N_b = 0$ as shown in upper figures. In the lower figures, there are floppy modes. In other words, when the coordination number is $N_v$, $N_b = 3N_v\text{-}6$ bonds are required to fix the shape of the polyhedron. Existence of the floppy part also means the location of free space in the structure. With the decreasing density of the system, a decrease of the coordination number occurs (Fig. 6.5a) and flexible-type structures also increase as shown in Fig. 6.12.

In the lithium silicate systems, $LiO_x$ polyhedra and networks formed by $SiO_4$ polyhedra are mixing. For the structures of porous lithium disilicate at $\rho = 1.98$, the coordination number of O around Si is 4 for ~100% of Si as found in the original disilicate glass at temperatures 600–800 K. That is, $SiO_4$ units are kept unchanged while the $N_V$ value of $LiO_x$ structure changes gradually. Therefore, structural changes in the porous systems are characterized well by the latter. The length of Si–O bonds (in Å) is only slightly changed with the density of the system. That is, the distance 1.493–1.846 is found for the original system, while 1.498–1.843 is found for $\rho = 1.98$ both at 600 K.

### 6.8.2 Changes in Coordination Polyhedra in Porous Systems

The concepts for the geometrical degrees of freedom in structures are applied to characterize the porous system as follows. As already shown in Fig. 6.5, each distribution of the coordination number $N_v$ (number of O atoms around Li ion) for each density of system at 800K was obtained for an instantaneous structure after the quasi-equilibration of the system. In this figure, the position of the maximum of the peak is found at 5 before the expansion of the system. The findings are comparable to the situation in lithium metasilicate, where the saturation of the total number of Li–O bonds and a comparable distribution were found [73] near $T_g$, when the temperature of the system was decreased. The increase of the coordination number with decreasing temperature means that the coordination polyhedra more overlapped each other at lower temperatures. In the case of $LiO_x$ polyhedra, the sum of $N_v$ values of all polyhedra corresponds to the total number of Li–O bonds. Therefore, this saturation means that further increase of overlap of polyhedra is not allowed. The peak position of the distribution shifts to 4 with a decreasing density as shown in Fig. 6.5a. The deviation naturally resulted in the enhanced diffusion of Li ions and this trend is following the relation between the slowing down of the dynamics near $T_g$ and the number of constraints discussed in [73]. The distribution patterns for 600 K (not shown) and 800 K are found to be almost the same and it means that the distribution mainly depends on density.

Original structure at 600 K is shown in the top panel of Fig. 6.12 and the porous system with $\rho$ = 1.98 (in g $cm^{-3}$) at the same temperature is shown in the bottom panel.

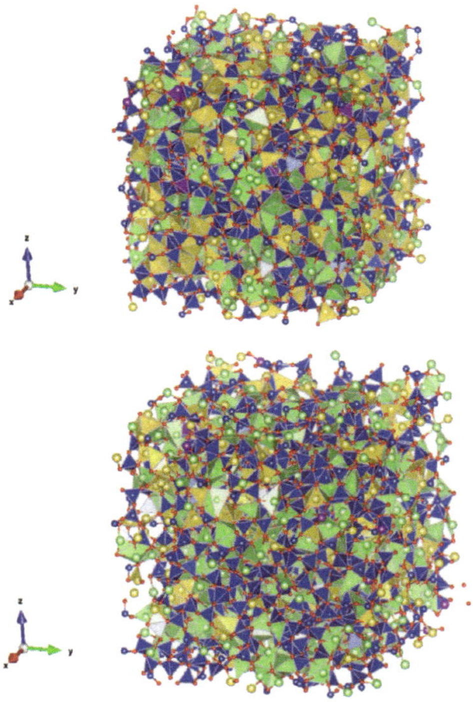

**Figure 6.12** Topological change of the coordination polyhedra (O atoms around Li ion) by introducing pores with decreasing density. Changes of the geometrical degree of freedom of the coordination polyhedra are shown for instantaneous structures at 600 K. Upper panel: Structure before expansion. The structure with $N_b = 3N_V$-6 is shown by yellow. The structure $N_{bc} < 3N_v$-6 (structure with a floppy mode) is shown in green. $SiO_4$ units are shown in blue. Lower panel: The structure after expansion at 600 K, for which the largest diffusion coefficient was observed ($\rho$ = 1.98). Colors are the same as those in the upper panel. By expansion, coordination polyhedra with $N_b = 3N_v$-6 decreased from 318 to 232 and the floppy structure with $N_b < 3N_v$-6 increased from 370 to 420. Here, total number of Li ions (and hence coordination polyhedra around Li ions) is 768. The size of each figure in these panels is proportional to the system size, approximately. Reprinted from Habasaki, J. (2016), Molecular dynamics study of nano-porous materials—Enhancement of mobility of Li ions in lithium disilicate in NVE conditions, *J. Chem. Phys.*, 145, 204503(1–11), with the permission of AIP Publishing.

In this figure, polyhedra with $N_b = 3N_V$-6 are colored yellow, while those with $N_b < 3N_V$-6 are colored green. The deformation of the coordination polyhedra is also affected by the $N_b$ explained below. The geometrical degree of freedom, $F_{polyhedron}$, changes by introducing pores. The degree is defined by $F_{polyhedron} = [(3N_v - 6) - N_b]$, where $N_v$ is the number of vertices, that is, the coordination number, and $N_b$ is the number of bonds (contact ion pairs for ionic liquid and contact oxygen pairs for $LiO_x$ structures). In the concept of deformation characterized by the $F_{polyhedron}$, the effects of bond-bending constraints are included. The $F_{polyhedron}$ is 0 when $N_b = 3N_V$-6, where 6 means the translational and rotational degrees of freedom of the polyhedron. In other words, $N_b$ braces characterized by this equality are required to fix the shape of the polyhedron. Thus, one needs to consider the change in $N_b$, which is relating to the mobility of the caged ions.

Coordination polyhedra for the network part ($SiO_4$ units) are colored blue. With expansion of the original system to this density, coordination polyhedra (yellow) with $N_b = 3N_V$-6 decreased from 318 to 232 (0.73 times larger) and the floppy structure (green) with $N_b < 3N_v$-6 increased from 370 to 420 (1.14 times larger), while the number of the structure with over-constraint ($N_b > 3N_v$-6) was almost unchanged. The number of polyhedra with a floppy mode ($N_b < 3N_V$-6) having a looser caging increases by the expansion of the system. This change, as well as that found in the coordination number of the cage of ions, are responsible for the changes in MSD. Thus, the porous systems are characterized by the increase of the structure with a floppy mode as well as smaller coordination numbers. Therefore, similar concepts as used to characterize the glass transition in the ionic liquids are found to be applicable to characterize the conduction, that is, "caged ion dynamics" in silicate glasses.

### 6.8.3 Effects of Tightening of Cages and Formation of Larger Voids

A decrease of the coordination number in $LiO_x$ polyhedra is found before the maximum of the diffusivity, while an increase of it is observed again after that. The latter change is accompanied with the rearrangement of networks after the formation of larger voids. Therefore, the maximum is explained by the change of the caging as a first approximation and we found that the existence of the larger void is not necessarily effective for the enhancement of the dynamics. These behaviors are closely related to the residual stress in the system and further study of the structures and stress in related composites will be interesting.

## 6.9 Relationships between Structure and Dynamics

Thus, the loosening of the cage is considered as the main factor to explain the enhancement of the dynamics observed in the porous materials in the *NVE* condition. Concerning these findings, some quantitative relations between the dynamics of ions and the changes in the coordination polyhedra were shown in previous works for lithium metasilicate glass. It was also shown that the spectra obtained for the fluctuation of $N_v$ and $N_b$ in the coordination polyhedra are closely related to those for the motion of caged ions in EMIM($NO_3$) [75].

In Fig. 6.13, partial MSD of Li ions is plotted for each $N_V$ value at initial times $t_0$ for the original and porous lithium disilicate systems ($\rho$ = 1.98) at 700 and 600 K. Here the partial MSD value for Li ions was obtained for 500 initial times covering 500 ps to 1 ns of time windows in each case. As shown in Figs. 6.13a,b, for the structure with smaller $N_v$ in the original system, the larger squared displacement of Li ions was found at around 1–4 ps. Jump motions of ions accompanied with the fluctuation of the coordination number started there. The initial increase of partial MSD depends on the coordination number $N_V$ and it decreases in the order of ($N_V$ = 2) > ($N_V$ = 3) > ($N_V$ = 4) > ($N_V$ = 5) > ($N_V$ = 6) > ($N_V$ = 7) both at 600 and 700 K. Li ions surrounded by two or three oxygen atoms show large mobility during several ps.

After the changing coordination number with time, behaviors are gradually averaged. While the partial MSDs in the case of $N_V$ larger than 4 are small and this trend continued for a long time. For the structure with $N_V$ = 6, the value in the short time scale was larger than that for $N_V$ = 5 and it decreased later. The value for $N_V$ = 3 also shows the decreases at longer time scales. The difference by the coordination number is larger at lower temperatures in the original system.

As shown in Figs. 6.13c,d, in the case of the porous system with $\rho$ = 1.98, the initial increases of the partial MSD with the fluctuation of the coordination number are observed much earlier time (~1 ps) as already discussed. The relation between the coordination number and the partial MSD is lost rapidly in the porous system. Nevertheless, a clear relationship between the coordination number and MSD exists but the coordination number alone is not enough to explain the behavior of ions in porous materials. One needs to consider the changes in $N_b$ values and changes in the time scale of the NCL region at the same time. A total number of bonds also play a role in the dynamics of the system.

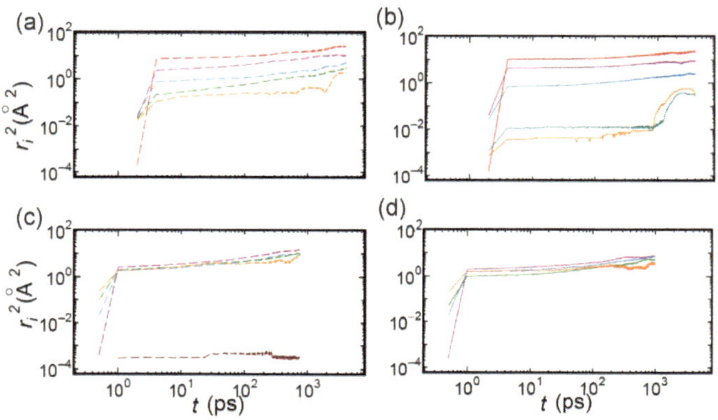


**Figure 6.13** Relation between partial MSDs of Li ions and the coordination number ($N_V$) determined at an initial time. These values are averaged for 500–1000 initial times. (a) The original system at 700 K. The $N_V$ values distribute from 2 to 6 (from upper to lower at the right part). Red: $N_V$ = 2, Pink: $N_V$ = 3, Blue: $N_V$ = 4, Green: $N_V$ = 5, Orange: $N_V$ = 6. The displacement of the system with a larger coordination number tends to be smaller. The difference by the initial coordination number is kept for a long time even after its gradual changes. (b) Original system at 600 K. $N_V$ changes from 2 to 6. Long-time localization was found especially for the polyhedra with the large coordination number. (c) Porous system with $\rho$ = 1.98 at 700 K. $N_V$ changes from 3 to 6 (from lower to upper at the beginning). The difference of mobility accompanied with the coordination number rapidly dissipated (by a change of the coordination number with the first jump). The result for $N_V$ = 7 (brown) is included here, although the numbers of events are not large enough for the statistical treatment. (d) Porous system with $\rho$ = 1.98 at 600 K. $N_v$ changes as similar manner as in (c) at the beginning. In porous systems, the structures for $LiO_x$ are rapidly changed by the jump to the next site. The same colors as in (a) are used in (b–d).

## 6.10 How to Separate Different Origins of Enhancement?

The methods used in the previous sections to separate the possible mechanisms of enhancement are summarized here.

Some of the procedures and concepts are applicable to examine other systems.

1. The systems with different porosity were prepared at 600 K. First, the effect of the difference of disorder and order in the structures was eliminated by using glass as a starting material as already mentioned.

   (Please note that the difference by ordering is quite large and should be considered on many occasions).
2. The systems at different temperatures with keeping the volume at each density (porosity).

   Thus, the volume change with temperature can be neglected here.
3. Mean-squared displacements of Li ions are examined for different densities.

   The behaviors in each time-region in porous systems are compared to clarify how events of each time region contribute to the enhancement.
4. Diffusion coefficients are plotted against density (porosity) and/or temperature.

The existence of the maximum in the former and slower slopes in the latter was confirmed.

To learn temporal aspects of dynamics, the following analyses were used.

1. To clarify the changes in time scale, each characteristic time in the mean-squared displacements of Li ions was examined [6, 7]. The contribution of the different mechanisms becomes clearer by it.
2. Since the enhancement already started in the early time (NCL or caging) region, this means the changes in caged ion dynamics being dominant for the enhancement. Then structures of coordination polyhedra were examined to learn the changes in the cage.

To specify the region characterizing the enhancement, the following analyses were used.

1. Direct observation of trajectories was done to determine spatial region where enhancement occurred.

   In the case of porous lithium disilicate examined for some slices, it was found that the enhancement is not necessarily along the boundary, although it does not exclude the possibility of the enhancement in layers with a larger thickness.
2. Structures of coordination polyhedra of Li ion were characterized by using the concept of the geometrical degrees of freedom of the polyhedra because the fluctuation of the cage and its decay are dominated by the degrees of freedom.

Enhancements of the dynamics are known to depend on the compositions, pore sizes, preparation methods and so on, and therefore the examination of mechanism in different conditions will be required, although one can expect some common mechanism underlying [8, 24].

In the present work, our attention was mainly focused on the single-phase system simulated in *NVE* conditions. However, the methods used to separate and characterize the enhanced dynamics would be expected to be useful for other systems with and without some modifications.

## 6.11 Comparison with Porous Lithium Disilicate and Other Porous Composites

In several nanoporous or mesoporous composite systems, the maximum of the transport properties observed in experiments tend to be explained by the percolation theory [31, 59–61].

For example, in Ref. [31], the existence of the maximum conductivity has been explained by the percolation theory with two critical points. In this (three-resistor) model, existence of enhanced interface conductivity is assumed. With increasing contents of insulator particles, the conduction paths percolate and further increase causes immersing of conductive materials in the insulator, which resulted in the decrease of conductivity (see Chapter 3 for more details).

On the other hand, in our results for porous the lithium disilicate systems in *NVE* conditions, the importance of the caged ion dynamics is found.

There are some differences between our systems and experimental nanocomposites discussed in Chapter 3, where enhanced dynamics have been observed. In several composites, conductivity maxima were obtained with increasing concentration of porous media (or insulator), while our observation of the maximum is diffusion coefficient plotted against density, so that only the effect of introducing pore is considered, without remarkable changing Li contents in *NVE* condition.

In the enhancement observed in composites, both changes of porosity and the content of Li ions occur at the same time. Therefore, the enhancement effect found in the experiment in such conditions exceeds the effect of decreasing the contents of mobile ions.

The difference in the thermo-dynamical conditions should also be taken into account because, in experiments, systems tend to be measured in *NPT* conditions.

Due to these differences, our results do not necessarily exclude other explanations of the enhancement for the composite immediately. For example, a layered structure of ions is worth to consider for effective conduction.

In spite of the limitation of the direct comparison, the methods used for the separation of the possible factors as shown in this chapter will be useful to understand the mechanism and applicable to other systems with more complex structures.

We also note here that insulator introduced in composites might have a role like pores introduced in *NVE* conditions. That is, it can modify the coordination polyhedra of ions.

Some cautions are required for the separation of possible origins of enhancement. Sometimes, limited numbers of factors and/or mechanisms are assumed from the beginning and one may overlook other possible causes. Furthermore, these possible factors of the mechanism are not necessarily separable completely.

Therefore, careful treatment of each problem is required, although the treatment of the system in this chapter is applicable to other systems.

## 6.12 Conclusion

In the porous lithium disilicate system, enhancement of the diffusion coefficient having the maximum in the medium density region is predicted by the molecular dynamics simulation in the *NVE* condition. In this chapter, procedures and concepts to separate the possible factors contributing to the enhancement were discussed and exemplified for the study of porous lithium silicate systems. The porous model system for MD is prepared by the expansion of the system started from the glassy lithium disilicate, so that the effect of disorder (difference of ionics in crystal and glass) was separated beforehand. The enhancement of the diffusive motion of Li ions is found to occur not only at the boundary of pores but also in the bulk part of the systems. The enhancement of the diffusion coefficient (and hence that of the conductivity) starts from an early time (NCL) region of the mean-squared displacement and this means that the loosening of the cage contributes to the enhancement. Changes found in the distribution of the coordination number and geometrical degrees of freedom of coordination polyhedra (formed by the oxygen atoms around Li ions) are consistent with this view.

However, our system is not a composite and the ratio of Li ions and Si atoms is kept unchanged, when the pore was introduced. Furthermore, in this chapter we discussed the simulation in *NVE* conditions not in *NPT* conditions. Therefore, the systems are not directly comparable to the Liang effect. Nevertheless, it is interesting to point out that the observed maximum in the medium density region of Li ion and smaller absolute slopes in Arrhenius plots are quite similar to those observed in composites of ionic conductors diluted by silica or alumina. We would like to suggest that the structural changes of $MO_x$ (M is mobile ions) or related polyhedra should be considered even in the composites and that the role of alumina or silica is not only the separator of paths, but also the modifier of the density of Li ions and geometrical structures of the cage.

As found in this chapter, nanoionics in porous materials is an interesting problem not only for nanotechnology but also for the fundamental research to understand the complex ion-dynamics and the relation to the glass transition problem. It provides a new platform to understand the heterogeneous dynamics in supercooled liquids and glasses.

# Appendix

## 6.A.1 MD Simulations and Modeling of Porous Systems

The system examined is lithium disilicate, which is located in the silica-rich region of $Li_2O$–$SiO_2$ systems, and pores are introduced by the scaling of volume and positions [13, 14]. So far, several compositions of lithium silicates such as lithium disilicate and lithium metasilicate in the molten, glassy, crystalline states have been examined successfully from many points of view by several groups, where our potential model previously developed based on the *ab initio* MO calculations [31] was commonly used. The model has been tested for several polymorphs of crystals under atmospheric pressure. The diffusivity and/or viscosity of the simulated system using our potential model are comparable to the experiments in several conditions [33]. Since the model covers a wide range of potential surfaces, it will be useful for further systematical comparison of systems with different states including the porous ones.

Each original and porous lithium disilicate ($Li_2Si_2O_5$) system examined contains 3456 atoms (768 Li ions, 768 Si atoms, and 1920 oxygen atoms) in the basic MD box with a periodic boundary condition. The following model function was used:

$$\phi_{ij} = \frac{z_i z_j e^2}{r} + f_0(b_i + b_j)\exp\left(\frac{a_i + a_j - r}{b_i + b_j}\right) - c_i c_j r^{-6} \tag{6.A.1}$$

The model potential consists of the Coulombic term, the pair potential function of the Gilbert–Ida type [76, 77] with the $r^{-6}$ for correction of the softness of the oxygen atom. The value $r$ is the distance between atoms and $a_i$ is the effective radius and $b_i$ is the softness parameters of the atom $i$ with a constant $f_0$ (=1 kcal Å$^{-1}$ mol$^{-1}$ = 4.184 kJ Å$^{-1}$ mol$^{-1}$). The Ewald method was used for the calculation of the Coulomnbic force. A cut-off distance was chosen to be 12 Å for the calculation of both the repulsive force and that for the real space term of the Coulombic force.

The present model seems to be good enough for discussing the trend caused by the changes in the density (and porosity) of the system and changes in dynamics in different states based on the same potential model.

### 6.A.2 Preparation of Porous Lithium Disilicate and Porous Lithium Metasilicate

The procedure for preparing porous systems is as follows. First, lithium disilicate in the glassy state was obtained by the rapid cooling (~1 K $ps^{-1}$) from the melt at 3000 K with the combination of constant pressure condition and temperature scaling to the target temperatures (2000, 1700, 1400, 1200, 1000, 800, 700, and 600 K). Systems were quasi-equilibrated at each temperature in the *NPT* condition and then the following runs in the *NVE* condition were done. Then porous systems were prepared by the scaling of the volume and position of particles at 600 K. Similar methods had been previously used to prepare the porous silica [13]. After the expansion of the system, each system was equilibrated during 300,000 steps run in the *NVT* condition and the following *NVE* runs are used for analyses. The time step used in the MD run was 1 fs for almost of porous systems. The densities $\rho$ examined are 2.47 (original), 2.30, 2.13, 1.98, 1.84 and 1.58 in g $cm^{-3}$ for lithium disilicate. Simulations at several temperatures (700, 800, and 1000 K) in the porous systems have been done using the same volume as systems at 600 K. The results for *NPT* conditions of MD are partly described in [6] and in Chapter 7. Interestingly, the self-healing process is found in this condition, even after the relatively large voids are formed. Because the porous system was prepared by the expansion of the original system in *NVE* conditions, observed enhancement may be correlated to the residual stress in the system. Therefore, the relation between such mechanical properties and dynamics in porous systems is also worthy to examine. The result in the subsequent *NPT* run related to this problem will be discussed in Chapter 7.

## References


1. Buurmans, I. L. C., Weckhuysen, B. M. (2012), Heterogeneities of individual catalyst particles in space and time as monitored by spectroscopy, *Nat. Chem.*, 4, pp. 873–878.
2. Madieh, S., Simone, Wilson, W., Mehra, D., Augsburge, L. (2007), Investigation of drug–porous adsorbent interactions in drug mixtures with selected porous adsorbents, *J. Pharm. Sci.*, 96, pp. 851–863.
3. Hara, M., Nakano, H., Dokko, K., Okuda, S., Kaeriyama, A., Kanamura, K. (2009), Fabrication of all solid-state lithium-ion batteries with three-dimensionally ordered composite electrode consisting of $Li_{0.35}La_{0.55}TiO_3$ and $LiMn_2O_4$, *J. Power Sources*, 189(1), p. 485.
4. Yu, Y., Gu, L., Zhu, C., Tsukimoto, S., van Aken, P. A., Maier, J. (2010), Reversible storage of lithium in silver-coated three-dimensional macroporous silicon, *Adv. Mater.*, 22, pp. 2247–2250.
5. Suib, S. L. (2008), Porous manganese oxide octahedral molecular sieves and octahedral layered materials, *Acc. Chem. Res.*, 41(4), pp. 479–487.
6. Habasaki, J. (2016), Molecular dynamics study of nano-porous materials—Enhancement of mobility of Li ions in lithium disilicate in NVE conditions, *J. Chem. Phys.*, 145, p. 204503.
7. Habasaki, J., Ngai, K. L. (2018), Heterogeneous dynamics in nano porous materials examined by molecular dynamics simulations-effects of modification of caged ion dynamics, *J. Non-Cryst. Solid.*, 498, pp. 364–371.
8. Habasaki, J., León, C., Ngai, K. L. (2017), *Dynamics of Glassy, Crystalline and Liquid Ionic Conductors: Experiments, Theories, Simulations*, 1st ed. (Springer International, Switzerland) and references therein.
9. Sing, K. S. W., Everett, D. H., Haul, R. A. W., Moscou, L., Pierotti, R. A., Rouquerol, J., Siemieniewska, T. (1985), Reporting physisorption data for gas solid systems with special reference to the determination of surface-area and porosity (recommendations 1984), *Pure Appl. Chem.*, 57, pp. 603–619.
10. Alothman, Z. A. (2012), A Review: Fundamental aspects of silicate mesoporous materials, *Materials*, 5, pp. 2874–2902; doi:10.3390/ma5122874.
11. Ma, L., Wen, Y., Yan, N. (2013), Pore size and channel structure study of mesoporous $SiO_2$ films on sensing trace explosive, *J. Sol-Gel*

*Sci. Technol.*, 67, pp, 161–169. https://doi.org/10.1007/s10971-013-3062-1.

12. Kelkar, S., Wolden, C. A. (2017), Evaluation of vapor deposition techniques for membrane pore size modification, *J. Membr. Sci. Res.*, 3, pp. 64–70.
13. Hampden-Smith, M. J., Kodas, T. T. (1995), Chemical vapor deposition of metals: Part 1. An overview of CVD processes, *Chem. Vapor Deposition*, 1(1), pp. 8–23. DOI10.1002/cvde.19950010103.
14. Kieffer, J., Angell, C. A. (1988), Generation of fractal structures by negative pressure rupturing of $SiO_2$ glass, *J. Non-Cryst. Solids*, 106 (1–3), pp. 336–342.
15. Nakano, A., Bi, L., Kalia, R. K., Vashishta, P. (1994), Molecular-dynamics study of the structural correlation of porous silica with use of a parallel computer, *Phys. Rev. B*, 49, pp. 9441–9452.
16. Beckers, V. L., De Leeuw, S. W. (2000), Molecular dynamics simulation of nanoporous silica, *J. Non-Cryst. Solids*, 261, pp. 87–100.
17. Bhattacharya, S., Kieffer, J. (2008), Molecular dynamics simulation study of growth regimes during polycondensation of silicic acid: From silica nanoparticles to porous gels, *J. Phys. Chem. C*, 112, pp. 1764–1771.
18. Habasaki, J., Ishikawa, M. (2014), Molecular dynamics study of coagulation in silica-nanocolloid–water–NaCl systems based on the atomistic model, *Phys. Chem. Chem. Phys.*, 16, pp. 24000–24017.
19. Rimsza, J. M., Du, J. (2014), Structural and mechanical properties of nano-porous silica, *J. Am. Ceram. Soc.*, 97, pp. 772–781.
20. Küchler, R., Kanert, O., Vereget, T., Jain, H. (2007), Effect of devitrification on ion motion in lithium-disilicate glass, *J. Non-Cryst. Solids*, 353, pp. 3940–3946.
21. Glass, A. M., Nassau, K., Negran, T. J. (1978), Ionic conductivity of quenched alkali niobate and tantalate glasses, *J. Appl. Phys.*, 49, pp. 4808–4811.
22. Minami, T., Takuma, Y., Tanaka, M. (1977), Superionic conducting glasses: glass formation and conductivity in the $AgI$-$Ag_2O$-$P_2O_5$ system, *J. Electrochem. Soc.*, 124, pp. 1659–1662.
23. Habasaki, J., Ngai, K. L. (2015), Molecular dynamics study of heterogeneous dynamics in lithium disilicate crystal, *J. Electroceram*, 34, pp. 43–56.
24. Tatsumisago, M., Machida, N., Minami, T. (1987), Mixed anion effect in conductivity of rapidly quenched $Li_4SiO_4$-$Li_3BO_3$ glasses, *J. Ceram. Soc. Jpn.*, 95, p. 197.

25. Ngai, K. L. (2011), *Relaxation and Diffusion in Complex Systems*, Springer, New York.
26. Maier, J. (1985), Space charge regions in solid two-phase systems and their conduction contribution—I. Conductance enhancement in the system ionic conductor-'inert' phase and application on AgC1: $Al_2O_3$ and AgC1:$SiO_2$, *J. Phys. Chem. Solids.*, 46, pp. 309–320.
27. Maier, J. (1985), Heterogeneous doping of silver bromide (AgBr: $Al_2O_3$), *Mater. Res. Bull.*, 20, pp. 383–392.
28. Sillassen, M., Eklund, P., Pryds, N., Johnson, E., Helmersson, U., Bøttiger, J. (2010), Low temperature superionic conductivity in strained yttria-stabilized zirconia, *Adv. Funct. Mater.*, 20, pp. 2071–2076.
29. Ohtomo, A., Muller, D. A., Grazul, J. L., Hwang, H. Y. (2002), Artificial charge-modulation in atomic-scale perovskite titanate superlattices, *Nature*, 419, pp. 378–380.
30. Thiel, S., Hammerl, G., Schmehl, A., Schneider, C. W., Mannhart, J. (2006), Tunable quasi-two-dimensional electron gases in oxide heterostructures, *Science*, 313, pp. 1942–1945.
31. Bunde, A., Kantelhardt, J. W. (2005), Diffusion and conduction in percolation systems, *Diffusion Condensed Matter*, pp. 895–914. DOI:10.1007/3-540-30970-5_22.
32. Habasaki, J., Okada, I. (1992), Molecular dynamics simulation of alkali silicates based on the quantum mechanical potential surfaces, *Mol. Sim.*, 9, pp. 319–326.
33. Habasaki, J., Okada, I., Hiwatari, Y. (1997), Fracton excitation and Lévy flight dynamics in silicate glasses, *Phys. Rev. B*, 55, pp. 6309–6315.
34. Gonçalves, L. G. V., Rino, J. P. (2014), Diffusion mechanisms in lithium disilicate melt by molecular dynamics simulation, *J. Non-Cryst. Solids*, 402, pp. 91–95.
35. Lammert, H., Heuer, A. (2010), Simplified interpretation of transport in disordered inorganic ion conductors from vacancy dynamics., *Phys. Rev. Lett.*, 104, p. 125901.
36. Mueller, C., Zienicke, E., Adams, S., Habasaki, J., Maass, P. (2007), Comparison of ion sites and diffusion paths in glasses obtained by molecular dynamics simulations and bond valence analysis, *Phys. Rev. B*, 75, p. 014203.
37. Montani, R. A., Balbuena, C., Frechero, M. A. (2012), Evidence of active regions for ion transport in lithium silicate glasses using the isoconfigurational ensemble, *Solid State Ionics*, 209–210, pp. 5–8.

38. Jabraoui, H., Vaills, Y., Hasnaoui, A., Badawi, M., Ouaskit, S. (2016), Effect of sodium oxide modifier on structural and elastic properties of silicate glass, *J. Phys. Chem. B*, 120, pp. 13193–13205.
39. Habasaki, J., Okada, I., Hiwatari, Y. (1995), Origins of the two-step relaxation and the boson peak in an alkali silicate glass studied by MD simulation, *Phys. Rev. E*, 52, pp. 2681–2687.
40. Shaw, R. R., Uhlmann, D. R. (1969), Effect of phase separation on the properties of simple glasses I. Density and molar volume, *J. Non-Cryst. Solids*, 1, pp. 474-498. and references therein.
41. Saha, D. R., Nandi, A. K., Chakravorty, D. (2013), Enhancement of ionic conductivity in $Li_2O$–$SiO_2$ glass in nanodimensions grown within pellets of ZnO nanorods and magnetodielectric properties of these nanocomposites, *J. Non-Cryst. Solids*, 376, pp. 12–17.
42. Liang, C. C. (1973), Conduction characteristics of the lithium iodide-aluminum oxide solid electrolytes, *J. Electrochem. Soc.*, 120, pp. 1289–1292.
43. Habasaki, J., Ngai, K. L., Hiwatari, Y. (2002), Molecular dynamics study of cage decay, near constant loss, and crossover to cooperative ion hopping in lithium metasilicate, *Phys. Rev. E*, 66, p. 021205.
44. Habasaki, J, Hiwatari, Y. (1999), Characteristics of slow and fast ion dynamics in a lithium metasilicate glass, *Phys. Rev. E*, 59, pp. 6962–6966.
45. Habasaki, J., Ngai, K. L. (2017), Nearly constant loss (NCL) in lithium metasilicate glass at low temperatures—anisotropic and dynamical caging from molecular dynamics simulations, *J. Phys. Chem. C*, 121(25), pp. 13729–13737.
46. Einstein, A. (1956), *Investigations on the Theory of Brownian Motion, Dover*, New York.
47. Kubo, R. (1957), Statistical-mechanical theory of irreversible processes. I. general theory and simple applications to magnetic and conduction problems, *J. Phys. Soc. Jpn.*, 12, pp. 570–585.
48. Odagaki, T., Lax., M. (1981), Coherent-medium approximation in the stochastic transport theory of random media, *Phys. Rev. B*, 24, pp. 5284–5294.
49. Haven, Y., Verkerk, B. (1965), Diffusion and electrical conductivity of sodium ions in sodium silicate glasses, *Phys. Chem. Glasses*, 6, pp. 38–45.
50. Jain, H., Peterson, N. L., Downing, H. L. (1983),Tracer diffusion and electrical conductivity in sodium-cesium silicate glasses, *J. Non-Cryst. Solids*, 55, pp. 283–300.

51. Nascimento, M. L. F., Fokin, V. M., Zanotto, E. D., Abyzov, A. S. (2011), Dynamic Processes in a silicate liquid from above melting to below the glass transition, *J. Chem. Phys.*, 135, 194703 (1–18), and references therein.

52. Tadanaga, K., Imai, K., Tatsumisago, M., Minami, T. (2002), Conductivity enhancement and thermal properties of AgI $MO_2$ (M = Zr, Si) composites using sol-gel derived aerogels, *J. Electrochem. Soc.*, 149, pp. A773–A777.

53. Mandelbrot, B. (1967), How long is the coast of Britain? statistical self-similarity and fractional dimension, *Science*, 156, pp. 637–638.

54. Ben-Avraham, D., Havlin, S. (2000), *Diffusion and Reactions in Fractals and disordered Systems*, Cambridge University Press.

55. Dräger, J., Bunde, A. (1996), Multifractal features of random walks and localized vibrational excitations on random fractals: Dependence on the averaging procedure, *Phys. Rev. E*, 54, pp. 4596–4602.

56. Bacry, E., Delour, J., Muzy, J. F. (2001), Multifractal random walk, *Phys. Rev. E*, 64, 026103(1–4).

57. Roman, E. H. (1992), Multifractality of random walks and localized excitations on linear random fractals, *Phys. A*, 191, pp. 379–385.

58. Kantelhardt, J. W., Zschiegner, S. A., Koscielny-Bunde, E., Havlin, S., Bunde, A., Stanley, H. E. (2002), Multifractal detrended fluctuation analysis of nonstationary time series, *Phys. A*, 316, pp. 87–114.

59. Telesca, L., Czechowski, Z., Lovallo, M. (2015), Multifractal analysis of time series generated by discrete ito equations, *Chaos*, 25, 063113; https://doi.org/10.1063/1.4922592.

60. Dieterich, W., Dürr, O., Pendzig, P., Bunde, A., Nitzan, A. (1999), Percolation concepts in solid state ionics, *Phys. A*, 15, pp. 229–237.

61. Roman, H. E., Bunde, A., Dieterich, W. (1986), Conductivity of dispersed ionic conductors: A percolation model with two critical points, *Phys. Rev. B*, 34, pp. 3439–3445.

62. Kul'kov, S. N., Buyakova, S. P., Smolin, A. Y., Roman, N. V., Kinelovskii, S. A. (2011), Percolation transitions in porous structure and their effect on physicochemical properties of ceramics, *Tech. Phys. Lett.*, 37, p. 360.

63. Iassonov, P. P., Beresnev, I. A. (2003), A model for enhanced fluid percolation in porous media by application of low-frequency elastic waves, *J. Geophys. Res.*, 108, B3, 2138 (1–8), doi:10.1029/2001JB000683.

64. Habasaki, J., Hiwatari, Y. (2002), Dynamical fluctuations in ion conducting glasses -slow and fast components in lithium metasilicate, *Phys. Rev. E*, 65, p. 021604.

65. Wang, B., Kuo, J., Bae, S. C., Granick, S. (2012), When Brownian diffusion is not Gaussian, *Nat. Mater.*, 2012, pp. 481–485.

66. Golding, I., Cox, E. C. (2006), Physical nature of bacterial cytoplasm, *Phys. Rev. Lett.*, 96, 098102(1–4).

67. Nakao, H. (2000), Multi-scaling properties of truncated Lévy flights, *Phys. Lett. A*, 266, pp. 282–289.

68. Mantegna, R. N., Stanley, H. E. (1994), Stochastic process with ultraslow convergence to a Gaussian: The truncated Lévy flight, *Phys. Rev. Lett.*, 73, pp. 2946–2949.

69. Koponen, I. (1995), Analytic approach to the problem of convergence of truncated Lévy flights towards the Gaussian stochastic process, *Phys. Rev. E*, 52, pp. 1197–1199.

70. Habasaki, J., Ngai, K. L. (2008), Molecular dynamics study of the dynamics near the Glass transition in ionic liquids, *Anal. Sci.*, 24, pp. 1321–1327.

71. Habasaki, J., Ngai, K. L. (2011), Multifractal nature of heterogeneous dynamics and structures in glass forming ionic liquids, *J. Non-Cryst. Solids.*, 357, pp. 446–453.

72. Habasaki, J., Hiwatari, Y. (2002), Molecular dynamics study of single and mixed alkali metasilicates-spatial and temporal characterization of the dynamics in the supercooled liquid and glassy states, *J. Non-Cryst. Solids*, 307–310, pp. 930–938.

73. Habasaki, *J.* (1990), Molecular dynamics study of the glass formation in $Li_2SiO_3$ system, *Mol. Phys.*, 70, pp. 513–528.

74. Loeb, A. L. (1976, 1991), *Space Structures* (Addison-Wesley, 1976; Birkhäuser, Boston).

75. Habasaki, J., Ngai, K. L. (2015), Rigidity and soft percolation in the glass transition of an atomistic model of ionic liquid, 1-ethyl-3-methyl imidazolium nitrate, from molecular dynamics simulations—existence of infinite overlapping networks in a fragile ionic liquid, *J. Chem. Phys.*, 142, 164501(1–19).

76. Gilbert, T. L. (1968), Soft-sphere model for closed-shell atoms and ions, *J. Chem. Phys.*, 49, pp. 2640–2641.

77. Ida, Y. (1976), Interionic repulsive force and compressibility of ions, *Phys. Earth Planet, Interiors*, 13, pp. 97–104.

Chapter 7

# Molecular Dynamics Simulations of Nanoporous Systems: Dynamic Heterogeneity, Self-Organization of Voids, and Self-Healing Processes

## 7.1 Introduction of Dynamic Heterogeneity in Porous Systems

Roles of porous materials are spreading in many fields of science and technologies, including nanoionics, and they are rapidly increasing in importance for both applications and fundamentals [1–3]. In this chapter, two topics concerned with porous systems will be discussed.

The first one is concerned with the dynamic heterogeneity of ionic motions in porous lithium disilicates. Heterogeneous dynamics meaning coexistence of fast and slow ions are commonly found in many systems such as glass-forming materials, ionically conducting glasses, colloidal systems, etc. [4]. To learn changes in heterogeneous dynamics caused by modifying their structures, porous materials are used. As discussed in Chapter 5, the dynamics

*Molecular Dynamics of Nanostructures and Nanoionics: Simulations in Complex Systems*
Junko Habasaki

ISBN 978-981-4800-77-8 (Hardcover), 978-1-003-04490-1 (eBook)
www.jennystanford.com

of particles (atoms, ions, molecules) show non-Gaussian characters in a certain time and temperature regions and the properties such as diffusion coefficients tend to be dominated by rare events of long-range. This is because the displacement of ion is squared in calculations of mean-squared displacement (MSD), which is connected to the diffusion coefficient through the Einstein equation [5]. Therefore, it seems to be interesting to learn how heterogeneous dynamics are affected by introduced pores and how they are related to the enhanced dynamics of porous systems discussed in Chapter 6. After summarizing the mechanism of the enhanced dynamics discussed in Chapter 6, our attention will be focused on the heterogeneity of the dynamics in ionically conducting porous glasses. The results are useful to learn the role of heterogeneity and how to control properties of materials by the perturbation.

The second one is concerned with the self-organization of larger voids in the *NVE* condition observed by the expansion of the system and healing processes observed in *NPT* runs after that. Results are useful to understand the effects of pressure and stress tensors related to them. Roles of extended conditions in molecular dynamics simulations and relation between structures are important factors to understand the dynamics toward to control nanostructures.

Now, let us start a discussion for the first topic. One may have many questions for the dynamic heterogeneity of porous materials. Are the heterogeneous dynamics also found in porous systems? How is it modified by introducing pores? As shown in Chapter 6, the enhanced dynamics of ions found in porous lithium disilicate in *NVE* conditions are accompanied with the structural changes of cages and paths of ionic motions. First, this system is examined to learn the relationships between changes of structures caused by introducing pores and those for dynamics, especially for exponents characterizing MSD. In the original glass before introducing pores, the dynamics of mobile ions are already heterogeneous. As shown in this chapter, enhanced dynamics found in porous lithium silicates obtained from MD simulations in *NVE* condition are also found to be heterogeneous ones [1]. Because the changes of dynamics are considered to be key features to characterize the slowing down of the dynamics in the glass transition problem, porous systems can provide a new platform to learn the mechanism of the glass transition and related problems.

In this chapter, density is given in g $cm^{-3}$ and time is in ps, unless otherwise stated.

## 7.2 Summary of the Enhanced Dynamics Found in Porous Lithium Disilicate

Before discussing the heterogeneity of the dynamics, we briefly summarize the essence of the results of Chapter 6 and [1, 2] for the enhanced dynamics found in porous lithium disilicate in *NVE* conditions. If one prepares porous systems started from crystalline ones, the difference between order and disorder in the structures will affect the results considerably. In our approach, possible effects caused by the different ordering in the system were eliminated by using glassy lithium disilicate as a starting material.

In *NVE* conditions, an increase of the diffusion coefficient of Li ions is found with decreasing density (increasing porosity). With a further decrease in density, the diffusion coefficient decreases again. That is, there is a maximum of the diffusion coefficient in the medium density region.

### 7.2.1 Enhancement Caused by Loosening of Cages

In Fig. 7.1, the structural changes caused by the expansion of systems are summarized by using mean coordination numbers. In Chapter 6, it was clarified that the increase of the diffusion coefficient of Li ions is caused by the loosening of the cage. This situation is also observed in Fig. 7.1.

The mean coordination number, $\overline{N}_V$, is found to decrease and then found to increase. This is consistent with the results observed in distributions of coordination numbers of oxygens around Li ions (Chapter 6). In this figure, wire frame structures are also shown with distinguishing rigidity of structures. Namely, the color of the bonds was changed by the geometrical degree of freedom [5] of $LiO_x$ structures.

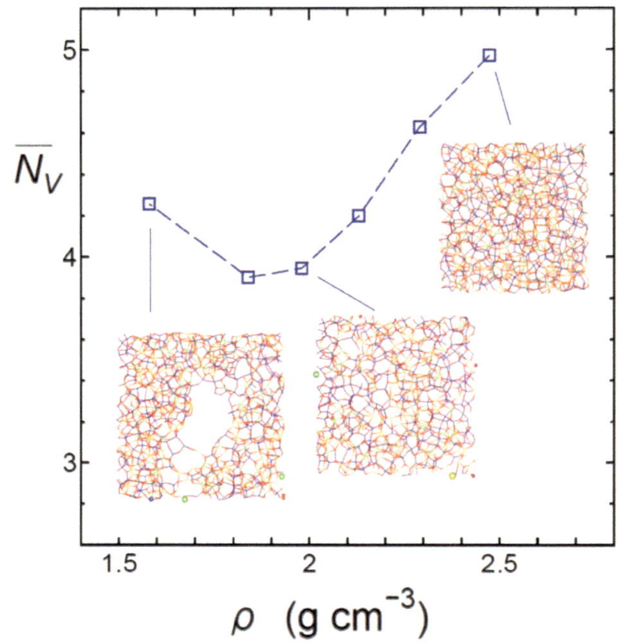


**Figure 7.1** Mean number of Li–O bonds around Li ion, $\overline{N}_V$, is plotted against density for the porous disilicate at 600 K. Cut-off length used is 3.0 Å. Original system before expansion is found on the right-hand side. The change in the mean coordination number is correlated well with the changes in dynamics. Namely, at around the minimum of the coordination number, the maximum of the diffusivity is found. Wire frame structures for the slice (The thickness of the slice is chosen to be 10 Å) are also shown. The number of contact pairs in O–O ($N_b$) is also related to the dynamics as well as the coordination number of O around Li ion ($N_V$). Colors of bonds change with the geometrical degree of freedom of the polyhedra. That is, the bonds for the structure $N_b<3N_V-6$ (with a floppy mode) is shown in green. Si part of the Si-O bonds in $SiO_4$ unit is shown in blue. Oxygen part of bonds is shown in red. With decreasing density, a loosening of the structure occurred and after that, a tightening of the structures occurred with formation of larger voids.

As discussed in Chapter 6, dynamics of ions are well correlated with such structural changes. With the decrease of the coordination numbers of $LiO_x$ structure, diffusion coefficient increases. The loosening of the cages in the small expansion region is also clear from the wire frame structures. With further expansion of the system, coordination number of $LiO_x$ structure increases. With this change, diffusion coefficient decreases, and it is related to the tightening of the cage. These tightenings coincide the formation of large voids as shown in this figure. That is, rearrangement of structure occurs with and after the formation of larger voids.

Thus, loosening of the cage causes the enhanced dynamics, while tightening of the cage after the formation of the larger voids causes the slowing down of the dynamics.

### 7.2.2 Explanation of the Mechanism by Geometrical Degrees of Freedom of Coordination Polyhedra

With changes of density, the maximum of the diffusivity appeared near the medium density (~2.0) with the system with enhanced dynamics (see Chapter 6). The trend in dynamics can be well explained by the changes in the coordination number; however, the contribution of other factors is non-negligible.

In general, the geometrical degrees of freedom [6] of the coordination polyhedra is determined by both the coordination number ($N_V$) and the number of contact pairs ($N_b$) of oxygen atoms and the latter also changes with density. That is, the change is characterized by both decreasing of coordination number and number of contact ion pairs. Corresponding changes in the time scale of the nearly constant loss (NCL) in susceptibility (i.e. caging) region in the MSD are found and therefore, the caged ion dynamics can be characterized based on the concept.

The concept of the geometrical degrees of freedom has been successfully applied for characterizing glass transition in several systems [7, 8]. These results suggest that explanation for the enhanced dynamics in porous materials is also relevant in discussing the mechanism of the slowing down of the dynamics by caging with decreasing temperature towards the glass transition temperature, $T_g$, in glass-formers.

In the following sections, it will be shown how enhanced dynamics is related to the heterogeneity and/or non-Gaussian parameters of dynamics.

## 7.3 Heterogeneous Dynamics of Ions in Porous Lithium Silicates

Recently, the importance of space and time heterogeneity in the catalyst was pointed out and several experimental methods begin to apply [9]. Under these circumstances, the knowledge obtained from the molecular dynamics simulation will be useful to understand the nature of dynamics in nanostructured materials. Here we will characterize the heterogeneous dynamics by MD simulations in porous systems [1]. In the original lithium silicate glasses, motions of ions were found to be extremely heterogeneous [3, 4] as shown in Chapter 5, and therefore it is interesting to examine how such heterogeneous dynamics are affected by introducing pores.

When we consider the effect of perturbation by introducing pores, it should be kept in mind that there are remarkable similarities in dynamics between ions in ionically conducting glasses, glass-forming molecules, and/or ionic liquids [3, 4]. This means that changes affected by such modifications of systems are important not only for understanding the ionics in ionically conducting glasses but also for understanding related slow dynamics in molecular and ionic glass-forming materials.

### 7.3.1 Non-Gaussian Parameters of Ions in the Porous System

Here the heterogeneity of the dynamics means the coexistence of fast and slow particles (ions) in transport properties [10]. How can we characterize the heterogeneity? One of the indexes to characterize the heterogeneous dynamics is the non-Gaussian (NG) parameters [11]. NG parameters, $\alpha_2$, are defined by

$$\alpha_2(t) = (3/5) <r^4(t)>/<r^2(t)>^2 - 1. \quad (7.1)$$

The parameter characterizes the deviation of $G_s(r, t)$ from the Gaussian form. That is, the NG parameter becomes 0 if the dynamics are Gaussian.

In Fig. 7.2, NG parameters of Li ions for the porous lithium disilicate obtained in *NVE* conditions for $\rho$ = 2.47 (purple, solid),

$\rho$ = 2.14 (pink, short-dashed), $\rho$ = 1.98 (red, 2-dot dashed) and 1.58 (blue, dashed) (densities are in g $cm^{-3}$) at 600 K are shown. In the bottom panel of Fig. 7.2, corresponding MSDs of Li ions are shown. For all porous systems examined, non-Gaussian characters are found at least for the time scales covered in the present work.

From this figure, one can see that at $\rho$ = 1.98, the non-Gaussianity is the smallest, although the heterogeneity cannot be neglected. As already shown in Chapter 6, at this density, the diffusion coefficient shows the maximum.

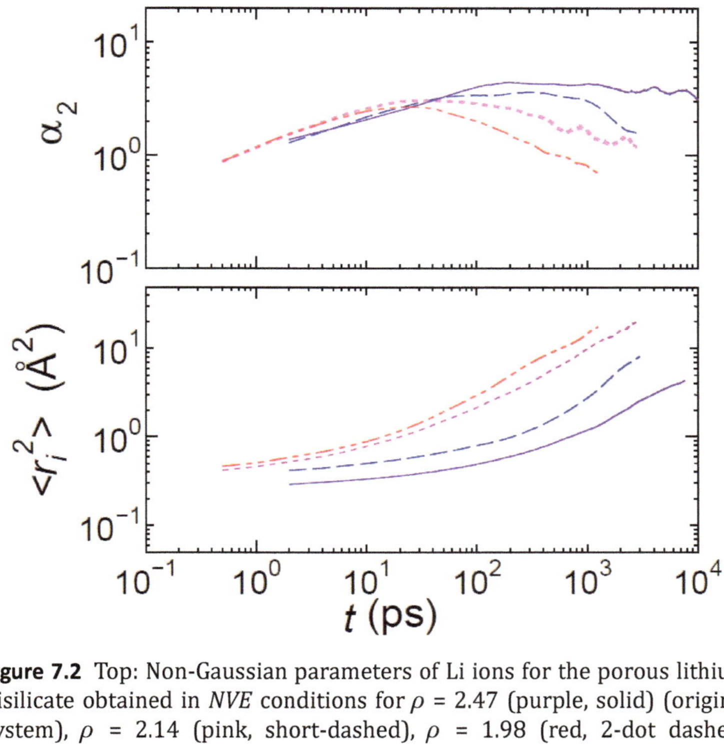


**Figure 7.2** Top: Non-Gaussian parameters of Li ions for the porous lithium disilicate obtained in *NVE* conditions for $\rho$ = 2.47 (purple, solid) (original system), $\rho$ = 2.14 (pink, short-dashed), $\rho$ = 1.98 (red, 2-dot dashed) and 1.58 (blue, dashed) at 600 K. In the porous system with the highest mobility, the non-Gaussianity decreases but still have a positive value for the longest time. Bottom: Corresponding mean-squared displacements of Li ions. The same kinds of curves and colors as those in the top column are used. The magnitude of the values shows the opposite trend to the Li ions. Reprinted with permission from Habasaki, J., Ngai, K. L. (2018), Heterogeneous dynamics in nano porous materials examined by molecular dynamics simulations-effects of modification of caged ion dynamics, *J. Non-Cryst. Solid*, 498, pp. 364–371. Copyright (2018), Elsevier.

If non-Gaussian parameters for porous systems and original system at 600 K were compared, the maxima of the parameter in porous systems are found to shift toward shorter time in the region with enhancement of dynamics [1]. The maxima are located near $t_{x2}$ [12], which is the beginning of the power law regimes after the jumping out of some of ions from their cages, and hence the value is correlated to the time, $t_{x1}$, when the caging region ends.

As discussed in Chapter 4, multifractality is another possible measure of the heterogeneity of structures and hence it will be interesting to apply multifractal analysis to the structures and dynamics of porous materials. Examples of such analyses are explained in Chapters 2 and 5.

## 7.4 Self-Part of the Van Hove Functions in Porous Lithium Disilicates

The self-part of the van Hove function is defined as follows [13]:

$$G_s(\mathbf{r},t) = (1/N)\sum_{i=1}^{N}\langle\delta(\mathbf{r}_i(t) - \mathbf{r}_i(0) - \mathbf{r})\rangle, \tag{7.2}$$

where $\mathbf{r}_i$ is the positional vector.

In Fig. 7.3a, the self-part of the van Hove functions at 600 K for $\rho$ = 1.98 is shown. As one can see, the functional form in Fig. 7.3a is different from the Gaussian form and has an inverse power law tails with exponential truncation. It can be regarded as the truncated Lévy distribution [14], although the bumps due to the jump characters of motions make the functional form more complicated. Both localizations within the site and motions with several forward correlated jumps still coexist for Li ions in porous systems, albeit with weaker caging.

If the function, $G_S(r, t)$, ($r$ = $|\mathbf{r}|$) obeys Gaussian form, the solution of the diffusion equation should be

$$G_s(r,t) = \frac{1}{(4\pi Dt)^{3/2}}\exp(-r^2/4Dt). \tag{7.3}$$

as shown in the example shown in Fig. 7.3b. In this figure, the maximum of the peak moves toward larger $r$ with time, while in Fig. 7.3a, the initial peak remains for a long time even when the fast ions reached the third and/or fourth shells. Although the width of the region of the tail part is rather limited in this case, the inverse power law tail (Lévy distribution) [14] with truncation [15–17] in an exponential form seems to be a good representation as found in other systems. The enhanced dynamics are dominated by long-ranged motions with higher jump rates and it is still extremely heterogeneous, although the NG parameter decreases considerably with the shortened time scales of jumps.

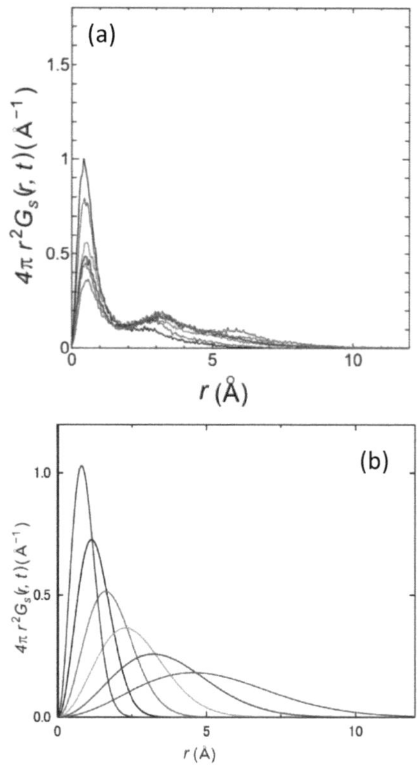


**Figure 7.3** (a) Self-part of the van Hove function of Li ions at 600 K for porous lithium disilicate at $\rho$ = 1.98, where the maximum diffusion coefficient was found. Curves are for 0.1, 0.2, 0.4, 0.6, 0.8 and 1.1 ns from left to right. (b) Example of a behavior expected from Gaussian dynamics, where $D$ = 2.0E$^{-11}$ (m$^2$ s$^{-1}$), $t$ = 0.08, 0.16, 0.32, 0.64, 1.28 and 2.56 (ns) were used. Reprinted with permission from Habasaki, J., Ngai, K. L. (2018), Heterogeneous dynamics in nano porous materials examined by molecular dynamics simulations-effects of modification of caged ion dynamics, *J. Non-Cryst. Solid.*, 498, pp. 364–371. Copyright (2018), Elsevier.

## 7.5 Percolative and Cooperative Aspect of the Dynamics

Time scales in porous lithium disilicate can be quantitatively examined by characteristic times of MSD for Li ions. As shown in Fig. 7.4 (Fig. 6.8 in Chapter 6 is modified slightly), the inverse of the characteristic times (in $ps^{-1}$) of Li ions in the porous lithium disilicate system obtained by the *NVE* condition shows the maximum at $\rho$ = 1.98 at 600 K and changes were already found to be at short time scales (in NCL region). In this chapter, the relation between the short-time-scale and long-time-scale motion will be discussed. The changes in caged ion dynamics explain the main features of the dynamics of ionics in porous systems at least in *NVE* conditions. However, can it explain all features? In the plot of Fig. 7.4, an asymmetry is found in a longer-time-scale behavior and larger change is found in the dense region as shown by a line with arrows. This means that the dynamics in porous systems cannot be explained by the changes in the caging alone. The time scale of this region is characterized by the power law exponent $\theta$ of the MSD before the diffusive regime. Changes in log $(t_{\mathrm{dif}}/t_{x1})$ against density at 600, 700, and 800 K are shown in Fig. 7.5a. These values are related to the time scale separation of caged ion dynamics and diffusive dynamics, which is governed by the geometrical correlations among jump motions after the caging region. In this figure, it was found that the minimum is clearer at higher temperatures. Thus, the changes in the trajectories of ions also contribute to the changes of the dynamics with density (and porosity). Another index to represent this separation of the time scale is the slope of the power law region of MSD, $\theta$. The values of $\theta$ plotted against density in porous lithium disilicate systems are shown in Fig. 7.5b for the same three temperatures. The maximum of $\theta$, which corresponds to the maximum of the diffusivity is found at 800 K, but clear maxima were not found for lower temperatures.

This situation can be explained as follows. Results in Figs. 7.5a,b are consistent with each other. The slope of the power law region of MSD is affected by the complexity of the trajectories, as explained in Chapters 4 and 5. In the power law region of MSD, both fast and slow ions are usually found in the original glass, and the slope $\theta$ is affected by the forward or back-correlated motion of jump motions of ions [3]. For the mean behavior of fast and slow ions, $\theta = 2/d_w$ is expected, where the exponent $\theta$ is a function of the fractal dimension of the random walk, $d_w$ [18, 19], which is an index to represent the complexity of trajectories.

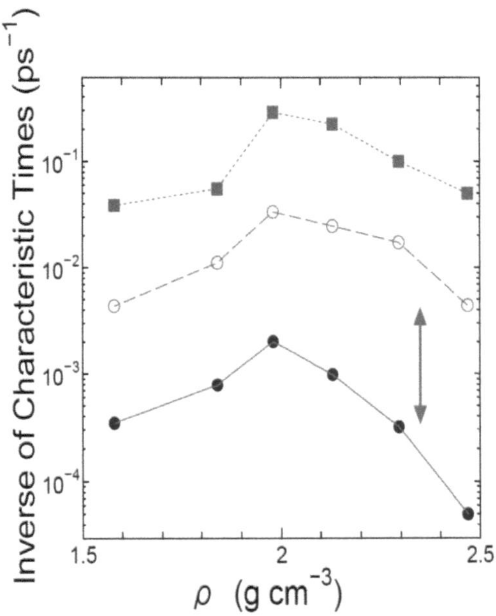


**Figure 7.4** Inverse of the characteristic times (in ps$^{-1}$) found in MSD of Li ions plotted against density of the system for lithium disilicate at 600 K. Filled squares, green: $1/t_{x1}$. Circles, blue: $1/t_{x2}$. Filled circles, dark blue: $1/t_{\text{dif}}$. Maxima are found even in the shortest time scale. The shape of the plot is asymmetrical especially for $1/t_{\text{dif}}$, because of the large difference of $1/t_{x2}$ and $1/t_{\text{dif}}$ in the right part of the plot (shown by a line with arrows). This means the large contribution of the change in the power law exponent, $\theta$, for the slowing down of the dynamics in the original system. Reprinted by permission from Habasaki, J., Ngai, K. L. (2018), Heterogeneous dynamics in nano porous materials examined by molecular dynamics simulations-effects of modification of caged ion dynamics, *J. Non-Cryst. Solid*, 498, pp. 364–371. Copyright (2018), Elsevier.

If the $d_w$ value is large, it means that the trajectory is a more complex one and ion tends to be localized. Therefore, the small value of $\rho$ (larger value of $d_w$) resulted in the retard of the time scale of $t_{dif}$. Thus, the large change in $1/t_{dif}$ was attributed to the geometrical changes of the trajectories, related to the structure of the jump paths and cooperative jumps. The result mentioned above for porous systems is useful to understand the dynamics in glasses, where ions (atoms, molecules) tend to be located near the bottom of the potential surfaces.

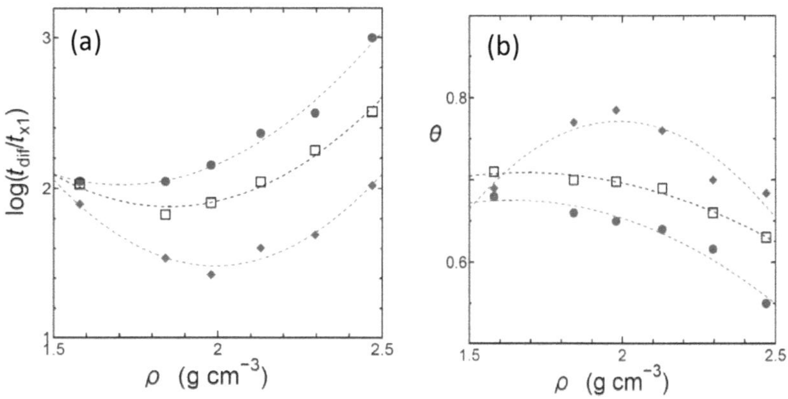


**Figure 7.5** (a) Temperature dependence of the value $\log(t_{dif}/t_{x1})$ plotted against density for three temperatures. 600 K: Blue, filled circles. 700 K: black, squares. 800 K: red, filled diamonds. (b) Temperature dependence of the power law exponent $\theta$ of MSD plotted against density. The same colors and marks as in (a) are used. The change in $\theta$ also contributes to the enhanced dynamics in the porous system and asymmetry of the behaviors in the diffusion coefficient plotted against density, but the contribution is smaller than the changes in the caging region. The maximum is found at high temperatures but not at lower temperatures. Reprinted with permission from Habasaki, J., Ngai, K. L. (2018), Heterogeneous dynamics in nano porous materials examined by molecular dynamics simulations effects of modification of caged ion dynamics, *J. Non-Cryst. Solid*, 498, pp. 364–371. Copyright (2018), Elsevier.

For the dynamics of ions in the glassy lithium disilicate system, change in the characteristic times related to the slope of the MSD in the power law region contributes to the slowing down of the dynamics considerably and this situation is changed by introducing pores. Thus, both changes in caged ion dynamics and power law exponent $\theta$ contribute to those in the dynamics in porous systems but differently.

The result suggests that power law exponent plays roles in the existence of the maximum at a high-temperature region as well as caged ion dynamics, while at low temperatures, changes in the caging dynamics are dominant for this. Results for the separation of different contributions of caging and power law exponent $\theta$ examined in the present work clarify the difference of dynamics of high-temperature and low-temperature regions.

The result discussed above is consistent with the previous results [20] found in an ionic liquid, EMIM (1-ethyl-3-methyl imidazolium)-$NO_3$. In this IL, diffusivity at a high-temperature region is found to be dominated by changes in the long-ranged motion, while that at low-temperature region, it is slowed down by the short-ranged motion (caged ion dynamics). This situation is comparable to the porous systems, although these contributions are mixing in some extents. In the porous systems, caging becomes looser and hence, fast dynamics motion observed at high-temperature region tends to continue at lower temperatures. As a result, the slope of the Arrhenius plot at the high temperatures is extended to lower temperatures. In the case of porous systems obtained by *NVE* conditions, the potential surface related to the longer $r$ region affects the motion of particles.

### 7.5.1 Explanation of Weakened Slowing Down of Ionic Motion with Decreasing Temperature in Porous Systems

The large decrease of $\theta$ was found in the original system with decreasing temperature (~0.55 at 600 K), and it resulted in a considerable increase in the time scale separation at low temperatures. The result means that the remarkable contribution of geometrical correlations in slowing down of the dynamics in the original glass.

The weaker temperature dependence of $\theta$ explains why the smaller absolute slopes in the Arrhenius plot of diffusion coefficients were found in porous systems [21]. That is, slowing down by complexity of trajectories is weakened in porous materials. The trend observed for $\theta$ in porous lithium disilicate can be explained by the following mechanism. In the lithium disilicate and related systems in the dense region, cooperative motions of several Li ions with long length scale with both forward correlated motion and back-correlated motions are found and the latter causes the slowing down of the dynamics at low temperatures. In porous systems, availability of the next vacant sites for jumps of ions increases and therefore it will decrease the back-correlation probability of ionic motions. The mechanism seems to be just opposite to the slowing down of the dynamics near $T_g$ but is related to the longer $r$ region of the potential functions. For an explanation of enhanced dynamics, both the changes in percolation of jump motions related to $\theta$ and those in caging dynamics for changes in dynamics should be considered.

## 7.6 Relation between Glass Transition and Changes of Slow Dynamics by Introducing Pores

As shown in Chapter 6 and Section 7.5, enhancement of dynamics in porous systems seems to be closely related to the mechanism of the glass transition, where both the path of particles and caging dynamics are changed. These results mean that one needs to consider both changes in percolation of jump motions related to the power law exponent, $\theta$, of MSD and those in caging dynamics for the enhancement, although the contribution of the former to the diffusion coefficient is larger in the case of the glass transition.

In the original system, fast ionic motion tends to be suppressed at lower temperatures, before ions being tightly caged. In the case of ionically conducting glasses, the behavior of these processes is not so deviated from the Arrhenius type temperature dependence, since the ionic motion is fairly decoupling from the motion of the matrix (that is, silica part, where packing of oxygen atoms play roles.) On the other hand, in the case of ionic liquids, ionic motion is directly related to the glass transition of the system and the fragile behavior, which can be represented as changes in slopes in the Arrhenius plot [20]. In both cases, a weakened cage is expected by introducing pores and therefore, the fast ionic motion will continue to lower temperatures.

The large effect of $\theta$ near the original glass means a remarkable contribution of geometrical correlations in slowing down of the dynamics in the original glass at the lower temperature. This conclusion is consistent with those in the study of the original system using fractal dimension analysis [3].

## 7.7 Self-Organization of Larger Voids

Heterogeneous dynamics mentioned in previous subchapters can be explained by the distribution of different structures in the environment of Li ions. In a short time scale, it is governed by the coordination polyhedra (and further shells) as already shown in Chapter 6. In the lowest density examined, gradually larger voids are formed, and it is interesting to examine this self-organization process of the void with the structure of LiOx.

First, the process of larger voids in the *NVE* condition will be re-examined from this point of view. In Fig. 7.6, time development of large void is shown by the wire frames and polyhedral structures of the system starting from the smallest density system at 600 K. In this figure, time development of structures of porous systems is shown along this formation process with the interval of 400 ps, where the polyhedra of $LiO_x$ are colored by its coordination number. On the left, many small pores are brought in the system. Different sizes of pores are found in systems. In the middle, the region containing the polyhedra with a small coordination number of Li developed and many small pores exist. In this stage, the system contains some structures having large coordination numbers, ~7. In the right panel, a larger void was formed with the rearrangement of polyhedra and networks. After the formation of the large void, contribution of large coordination numbers becomes negligibly small and the coordination number of 4–6 becomes dominant.

This structure with the larger void was essentially kept after further 400 ps, while the further rearrangement of coordination polyhedra is found. The porosity of the system is a linear function of the density (see, Eq. (6.1) of Chapter 6 or Appendix of this chapter) even after the formation of larger voids, and this means that it accompanied with the decrease of free space (small pores) in other parts. Due to rearrangements of the rest part, the smaller pores found in the left and middle panels disappeared. That is, a rearrangement of the structure occurs with keeping total free volume.

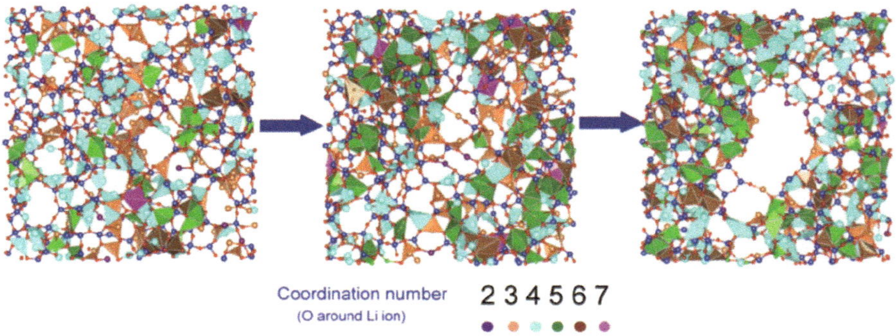


**Figure 7.6** Development of large void in the system at $\rho$ = 1.58 with an elapse of time from left to right, which is shown in the interval of 400 ps (100,000 steps) at 600 K for the slice of 10 Å. Each polyhedron (Li ions surrounded by oxygen atoms) is colored by its coordination number, $N_V$, shown in the figure. Blue spheres: Si. Red spheres: O. Pale blue spheres: Li. Li ions colored purple are for the structure with $N_V$ = 2.

The size of the larger void discussed here is about ~2 nm in diameter for the case observed in Fig. 7.6. Based on the recommendation by IUPAC for the field of physisorption [22], the classification of porous systems is mentioned as follows. "In the context of physisorption, it is expedient to classify pores according to their sizes: (i) pores with widths exceeding about 50 nm are called macropores; (ii) pores of widths between 2 and 50 nm are called mesopores; (iii) pores with widths not exceeding about 2 nm are called micropores." Even the large void found in the present simulations is still between microporous and mesoporous regions. Because the porous system in the present work was prepared by the expansion of the original system in *NVE* conditions, observed enhancement is correlated with the residual stress in the system.

These effects of the residual stress in the system are comparable to that found in the stress–strain curve [2] because pressure gradually changes during this process.

## 7.8 Effect of Stress Tensor on the Enhancement

In Fig. 7.7, pressures of the system obtained in *NVE* conditions at 600 K in the quasi-equilibrate states are shown as a function of the density.

The minimum is found near the maximum of the diffusion coefficient. This means that the loosening of the cage in the porous lithium disilicate is accompanied with the large negative pressure of the system in *NVE* conditions. In other words, the enhancement is suggested to be caused by the changes in the stress tensor in the system.

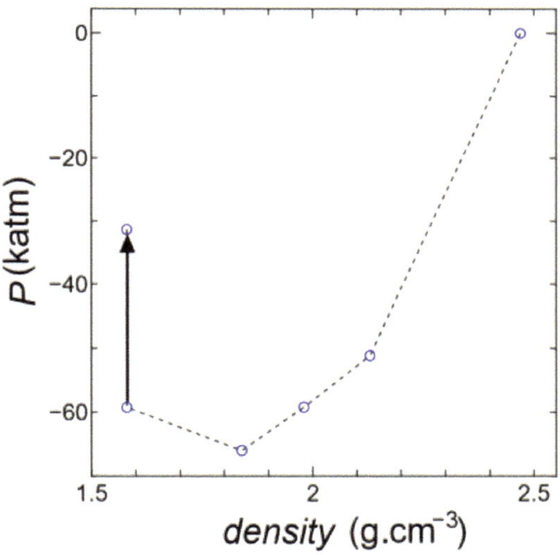


**Figure 7.7** The pressure of the system under *NVE* conditions plotted against the density of the system. At the lowest density in the plot, the negative pressure is found but it increased gradually (see arrow) with the formation of a larger void observed in Fig. 7.6.

In the case of the lowest density ($\rho$ = 1.58), an increase of the pressure was observed during the formation of larger voids in *NVE* condition as shown by the arrow in the figure. This means that the further rearrangement of the structure was brought by the stress relaxation in the system. In the following *NPT* run, shrinks of the size of voids are observed with increase of the pressure (to ~0 atm). However, voids formed do not disappear even after 1 ns (1,000,000 steps) run. The result of a longer length scale will be shown later.

It is useful to see how the pressure of the system is related to this behavior. The pressure of the system is defined as follows from the virial theorem [23]:

$$P = \frac{1}{V}Nk_BT + \frac{1}{3V}\left\langle \sum_{i=1}^{N} \boldsymbol{r}_i \cdot \boldsymbol{F}_i \right\rangle$$

$$= \frac{1}{V}Nk_BT - \frac{1}{6V}\left\langle \sum_{i=1}^{N} \sum_{\substack{j=1 \\ i \neq j}}^{N} r_{ij} \frac{d}{dr_{ij}} \varphi(r_{ij}) \right\rangle, \qquad (7.4)$$

where $r_{ij} = |\boldsymbol{r}_i - \boldsymbol{r}_j|$.

Accordingly, the pressure of the system, as well as each component of the pressure tensor of the system, is a function of the position of the particles on the potential surface.

To consider the mechanism of the enhancement in other systems such as nanocomposites discussed in Chapter 3, it seems to be important to examine the stress tensor and related structures. It will be interesting to examine if the a similar structure and/or mechanical changes occurred in other systems showing enhanced dynamics and if similar structural changes without negative pressure can cause the enhancement or not.

Thus, the mechanism of the enhancement is closely related to the mechanical properties found in the stress–strain curves and nanofracture of the system. In the stress tensor, its components are stresses exerted across surfaces perpendicular to the coordinate directions. The observed relation between pressure and volume in Fig. 7.7 is comparable to the relationship between stress and strain if we considered the difference of plus and minus sign of values and dimensionality.

Recently, Wang et al. [24] examined the stress–strain curve of sodium silicate and silica using MD simulations. As shown in Fig. 7.8, the silicate system shows ductile behaviors and after that, nanofracture is observed. The result is compared with that of silica, where brittle behavior is observed. In the sodium silicate, the following observation was reported [24]. "As shown in the inset of Fig. 7.8, as the strain increases, the relative deformation of Si–O bonds presents the same shape as that of the stress–strain curve and eventually goes back to its initial zero-stress value.

This means that Si–O bonds essentially deform in a reversible elastic fashion under stress. On the other hand, Na–O bonds behave in a significantly different way. Indeed, the maximum relative elongation of the Na–O bonds is much lower than that of the Si–O bonds, which shows that, after a short elastic regime, these bonds yield at low stress and, thereby, initiate the fracture through some plastic deformations prior to the failure of any Si–O bond."

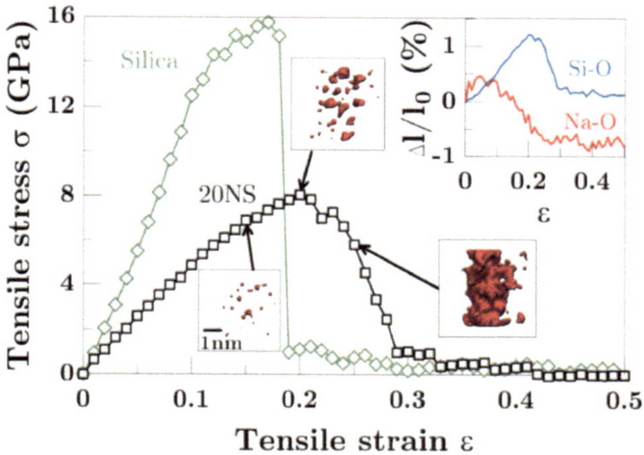


**Figure 7.8** Stress-strain response of the simulated $(Na_2O)_{20}(SiO_2)_{80}$ glass during fracture. The snapshots illustrate the volume of the cavities, with a radius larger than 5 Å, that are formed at various strains. The inset shows the relative variation of the average Si–O and Na–O bond lengths ($\Delta l/l_0$) with respect to the strain. The cut-offs used to identify Si–O and Na–O bonds are 1.974 and 3.311 Å, respectively, as determined from the position of the first minimum after the peak associated to the first coordination shell in the pair distribution functions. Reprinted from Wang, B., Yu, Y., Wang M., Mauro, J. C., Bauchy, M., *Phys. Rev. B*, 93, 064202 (2016) with permission.

From these results, alkali metal ions seem to take important roles in the mechanical properties of the system. In the case of porous silicates, the distances between pairs of atoms (or ions) are shown as the function of the density in Fig. 7.9a. For the distance in each pair, a position of the maximum of the first peak in $g(r)$ is used. In this plot, there are no remarkable changes in Li–O and Si–O pairs. Instead, changes in the Li–Li distance seem to be large and correlated well for the changes in the pressure of the system.

The position of the maximum for the Li–Li distance is also found near the maximum of the diffusivity. A small maximum is also observed for Li-Si and Si-Si pairs. These results mean that the changes in the structures are mainly for spaces without direct bonding. Therefore, structural changes related to these parts should not be neglected for the arguments of mechanical behaviors.

At first sight, this result is not the same as those in the inset of Fig. 7.8. However, this result is not necessarily against the result concerned with the local structures, because the changes shown in the inset are in an order of ~1% in length. To examine the behaviors of this order, in Fig. 7.9b, the values for Si–O and Li–O are replotted in an enlarged vertical axis. Then, similar trends as in Na–O and Si–O bonds in the inset of Fig. 7.8 are found for Li–O and Si–O. As shown in this figure, the changes in Li–O occurred at the beginning of the expansion and those for Si–O occurred later. The trend observed is consistent with that in sodium silicate.

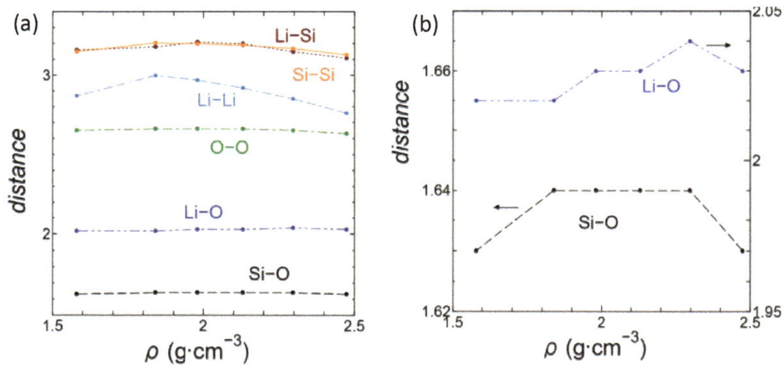


**Figure 7.9** (a) Density dependence of the distance between neighboring atoms for all pairs. The distance for Li–Li ions shows the largest change with density in our model. It becomes longer with decreasing density. With the formation of voids, it returns to be short. For Li–Si and Si–Si pairs, maxima are also observed. (b) Density dependence of the distance for Si–O and Li–O pairs in enlarged scales. The distance for Li–O increases with decreasing density and it decreases with a further decrease of density. The maximum for Si–O is not so clear but seems to be located in a smaller density region than in Li–O. Trends are consistent with the results in the inset of Fig. 7.8.

From their results, it is clear that alkali metal ions take important roles in the mechanical properties of the system.

That is, they prevent the brittle behavior of the system and make rearrangement of the system easier.

This finding is important to consider the self-healing process of the system in the following sections.

### 7.8.1 Self-Healing Process in NPT Conditions

The system prepared in the constant-volume condition is not necessarily stable under atmospheric pressure. The system will shrink because of negative pressures in the system equilibrated in *NVE* conditions when the porous system is prepared by expansion.

This process seems to be interesting from the following two points of view; one is controlling the properties of porous materials and another is related to the mechanism of the self-healing process that occurs in porous systems. They are also related to the mechanism of nanofractures. Here, we will examine porous lithium disilicate and metasilicate systems from these points of view.

### 7.8.2 Self-Healing Process in Porous Lithium Disilicate

If molecular dynamics simulation of porous lithium disilicate (see Chapters 6) obtained in the *NVE* condition was followed by *NPT* runs, recovery of the system density was observed in several cases.

In Fig. 7.10, the recovery curves of density starting from the $\rho$ = 1.98 are shown. During the *NPT* run, the system gradually shrank to $\rho$ = 2.44 and the pressure of the system decreased to ~0 k atm within 150 ps at 600 K. At this density, large voids are not formed, and the system can easily return to the original density when the *NPT* condition was applied. The data of runs at different temperatures are shown in the same figure. Even if the *NPT* run is performed at 300 K after the *NVE* run at 600 K, the result is comparable to the run at 600 K. In the case of the run at 300 K after the quench at 100 K, the results are also quite similar. For the data at 600 K, some data points in shorter time scales are shown. As shown in this figure, a large fluctuation of density occurs on a short time scale. These results suggest that the process observed in this region is not controlled by thermally activated processes such as diffusive motions. It seems to be caused by the ductile behavior of the system and/or with fluctuations of structures in a short length scale. The existence of the negative pressure of the system means that the process is related to the stress relaxation and mechanical properties of the system.

Our results are consistent with those by Wang et al. [24] for sodium silicate. That is, in the small- and medium-density regions, the system of alkali silicate shows an elastic property and nanoductility, respectively. Formation of larger voids observed in the lower the density region ($\rho \leq 1.58$) seems to correspond to the brittle behavior (nanofracture) of the system found in sodium silicate in [24].

In Fig. 7.11, snapshots of ball-and-stick structures before and after self-healing processes in *NPT* run of lithium disilicate at 600 K, where the run was starting from $\rho$ = 1.84, are shown. In this case, recovery is almost completed within ~10 ps. This time scale seems to be quite short, where ions are still in the local region, judging from the MSD of several temperatures shown in Fig. 7.2. Therefore, essentially, this process is considered not accompanied with diffusive motion, although there is a non-zero probability of diffusive motions at this temperature.

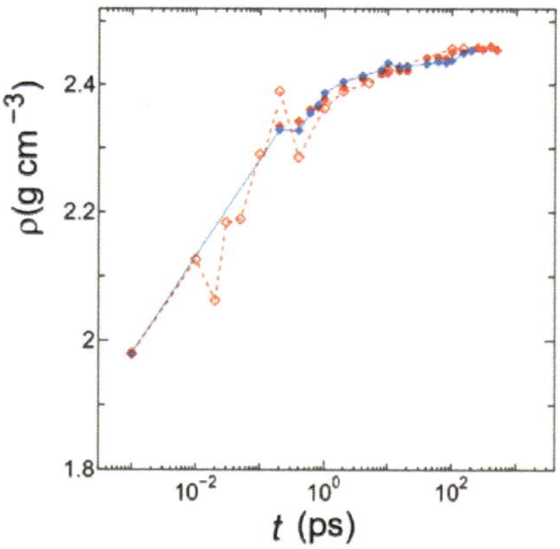


**Figure 7.10** Self-healing processes observed in porous lithium disilicate systems during *NPT* runs at 600 K (blue, open diamond) and at 300 K (red, filled diamond) starting from $\rho$ = 1.98. The result at 300 K after the quench at 100 K is also shown (pale blue, filled diamond). Data are connected for guiding the eyes. When we started the comparable density, the resultant curves are similar even though different temperature conditions are used. That is, the process is not governed by the thermally activated processes in this case.

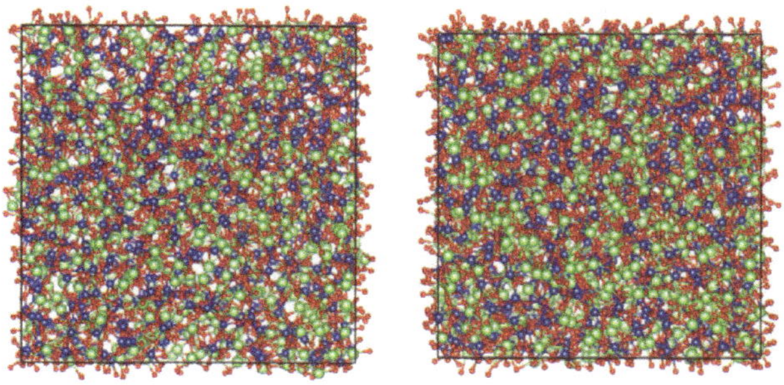

**Figure 7.11** Example of the healing process in structures of porous lithium disilicate at 600 K. Before (left) and after (right) self-healing processes in *NPT* run. The run was starting from $\rho$ = 1.84. In this case, almost complete recovery of the density was observed within ~10 ps run. Blue: Si atoms. Red: O atoms. Green: Li ions. The size of each figure is nearly proportional to the side length of MD cell of each system.

As shown in Fig. 7.12, when the *NPT* run is started from the density, $\rho$ = 1.58, the void remained after ~2 ns. In this case, fractured surfaces are separated from other parts and in each part, tight network structures exist. Initiation of healing probably requires longer lead times to wait for the suitable geometrical conditions of atoms being satisfied. Furthermore, diffusivity the of ions is slower than the medium density region, although the motion of atoms is of a longer length scale required for the recovery of the atoms separated from other regions. This explains why the process is slower than in other cases. Even in this case, once the density becomes larger, the acceleration of the process occurs as found in the change of the slope at around 100 ps.

The recovery of density for the above two cases is compared in Fig. 7.13.

In the case of the starting from $\rho$ = 1.84, a rapid recovery occurred at around 0.2~1 ps. When it started from $\rho$ = 1.58, a rapid increase in the density was not observed even at around $\rho \sim 1.84$.

It is interesting to note that fluctuations are observed in the long-time region in the density curves. This means the initiation and termination of the healing process are alternated, which suggests that the system needs a suitable arrangement for starting of healing processes. It may also be related to the heterogeneity of the structures and dynamics discussed in the first half of this chapter.

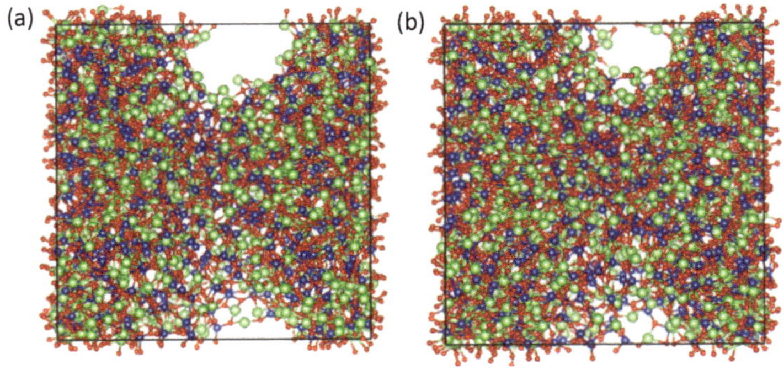


**Figure 7.12** Example of (a) before and (b) after self-healing processes in *NPT* run. The system is lithium disilicate at 600 K started from $\rho$ = 1.58. Complete recovery of the density was not observed after ~2 ns. Pressure of the system becomes ~0 in the earlier stage.

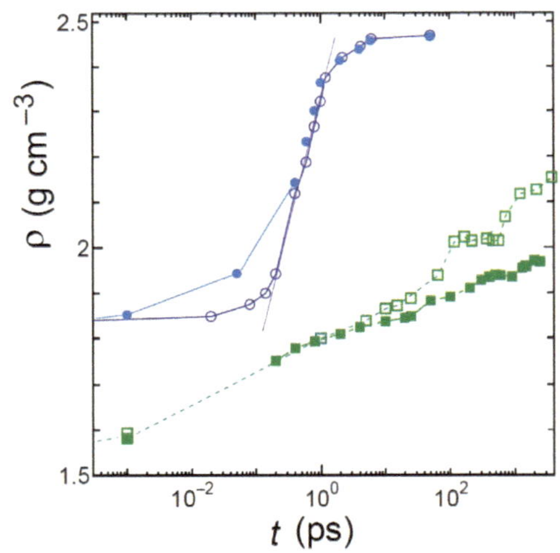


**Figure 7.13** Upper curves: examples of self-healing processes observed in porous lithium disilicate systems during *NPT* runs starting from $\rho$ = 1.84. In the *NPT* runs at 600 K (blue, open circles) and at 300 K (blue, filled circles). A solid line is a fitted one with a logarithmic function. Power law functions are also applicable. Lower curves: Examples of self-healing of the same system during *NPT* runs at 600 K (green, open squares) and at 300 K (green, filled squares) starting from at lower density $\rho$ = 1.58, where the system shows a larger void in the *NVE* condition. The healing process is much slower and the difference between at 300 K and 600 K is remarkable at a longer time region than ~1 ps. A large fluctuation of the curves was observed at lower curves especially at 600 K. Data points are connected for guiding the eyes.

Different from the healing at a higher density region, if the run started from the lower density, the process is found to be temperature-dependent. As shown in Fig. 7.13, the difference between the curves at 600 and 300 K is found after 1 ps and becomes larger at longer times. The importance of the activated process such as jump motion of ions for the initiation is suggested by this fact. In porous lithium disilicate systems starting from lower density, some holes tend to remain in the systems even after the longer runs. At the longest time (~4 ns), the density already exceeds 2.15, where the rapid healing process is observed if one started the *NPT* run from that density after the expansion. Therefore, the slow increase of the density means the stability of the system with these holes after rearrangements of networks. This is caused by the formation of rigid structures of holes being separated from other domains. The healing process is much slower than that observed at the case starting from higher density and the difference between at 300 and 600 K is remarkable at a longer time region than ~1 ps. A large fluctuation of the curves is observed at lower curves especially at 600 K.

Further recovery of the system will not be easy once the ~0 pressure condition was achieved.

In Fig. 7.13, a fitted line using a logarithmic function is shown for data at 600 K starting from the higher density. In this case, the curve in the large change in the middle part can be fitted to

$$\rho = 0.232\ln t + 2.32 \tag{7.5}$$

The same data can be fitted to the power law as well:

$$\rho = 2.32t^{0.109} \tag{7.6}$$

It is interesting to note that the power law is also observed in the fracture of materials [25, 26] in several kinds of plots.

From the lower density curves with large fluctuations of the data, it is not clear what functional form is suitable, although these functions, as well as a linear function, can be applied.

### 7.8.3 Self-healing Process in Porous Lithium Metasilicate

Porous systems can be similarly prepared for lithium metasilicate as in lithium disilicate [27]. In lithium metasilicate with higher alkali contents, the self-healing process is also observed, and the recovery of the system is found to be almost completed within the relatively short observation time. It is found even after the larger void being formed in the *NVE* run as shown in Figs. 7.14 and 7.15. Naturally, a larger number of Li ions seem to make the system easier to cause the rearrangement of the structure, although further examinations are required to conclude it. Furthermore, the diffusivity of Li ions in metasilicate is larger than the disilicate system [3].

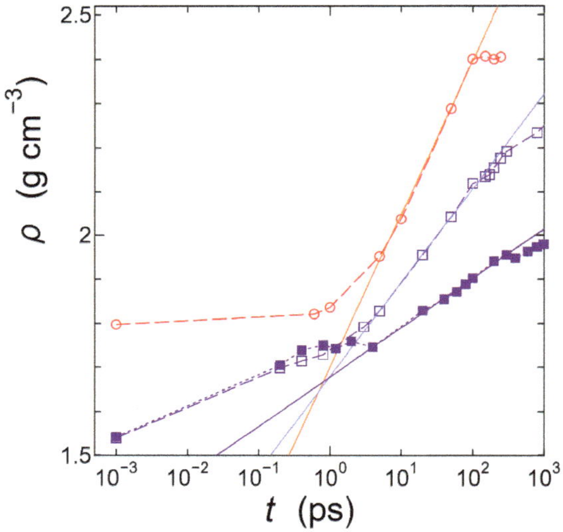


**Figure 7.14** An upper curve: examples of self-healing processes observed in porous lithium metasilicate systems during *NPT* runs starting from $\rho$ = 1.80 in the *NPT* run at 600 K (red, open circles). Lower curves: Examples of self-healing systems of the same system during *NPT* runs at 600 K (blue, open squares) and at 300 K (blue, filled squares) starting from at lower density $\rho$ = 1.53, where the system shows a formation of larger voids in the *NVE* condition. The healing process is slower than that starting from higher density and the difference between at 300 and 600 K is remarkable in a long-time region ~1 ps. Data points are connected for a guide of eyes and fitted lines to logarithmic functions are shown.

In Fig. 7.14, linear parts can be seen in the plot of density against logarithmic of time in both systems starting from $\rho$ = 1.53 and 1.80 at 600 K.

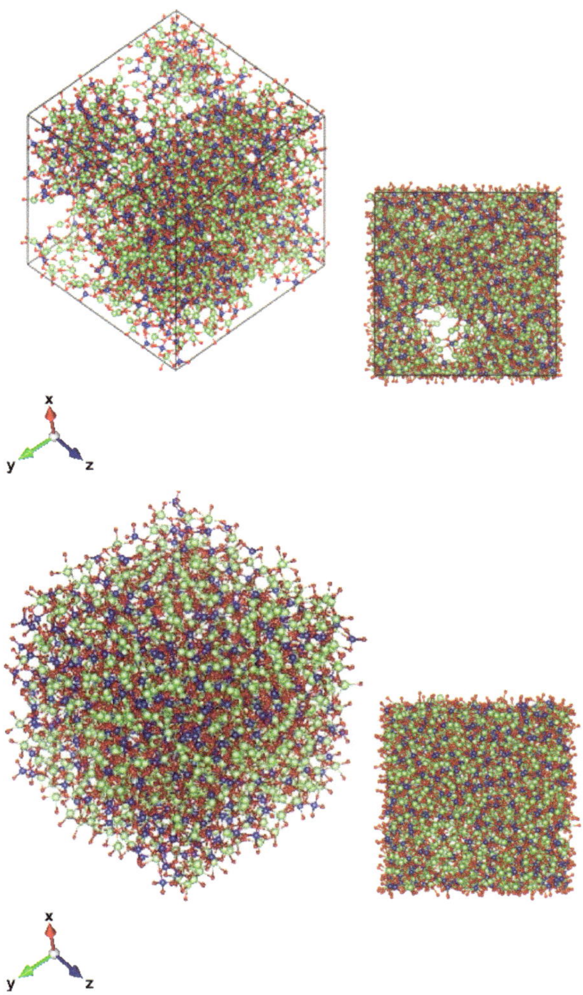


**Figure 7.15** Self-healing processes observed in porous lithium metasilicates. Example of (a) before and (b) after self-healing processes in *NPT* run for lithium metasilicate at 600 K starting from $\rho$ = 1.53. In this case, almost complete recovery of the density was observed within ~2 ns run.

The fitted lines for the former and latter, respectively, are

$$\rho = 0.150 \ln t + 1.70 \tag{7.7}$$

and

$$\rho = 9.29\text{e}{-2}\ \ln t + 1.68 \tag{7.8}$$

The result at 300 K starting from the lower density is also shown in the figure. After several ps, the deviation from the curve at 600 K becomes larger and it again confirms that the process is affected by the thermally activated process. In the linear part of the log scale plot, a relation

$$\rho = 3.24\text{e} - 2\ \ln t + 1.75 \tag{7.9}$$

is found.

In all three cases mentioned above, power laws are also good approximations as in the porous lithium disilicate system.

The process depends on temperatures and initial density; however, the existence of the common intersections (1.68–1.75) suggests that systems have both the core part and the flexible part sensitive for repairing.

The process in the final stage after the log or power-law region is slower than early processes. Different from the case of disilicate system, the fluctuation of the density during the healing process in the lithium metasilicate system starting from the low density with large voids ($\rho$ = 1.53) is not remarkable (see Fig. 7.14). The process seems to proceed rather smoothly, and it seems to be accelerated by a larger number of Li ions. Even in the case with larger voids, the repairing of the systems is almost completed as shown in Fig. 7.15, within ~2 ns.

These processes consist of several regions of different mechanisms. As shown in Fig. 7.16, the difference in the results at higher-density and lower-density regions is observed for the time dependence of the pressure during the *NPT* runs. Although the process might be affected by the time constant and related weight of the wall of the MD, pressures were controlled by the Nosé-Hoover method with the same time constant of 1 ps.

As shown in Fig. 7.16, in the case of the run starting from $\rho$ = 1.80, pressure gradually increases during the healing process until ~200 ps, which corresponds to the end of the logarithmic (or power law) region. Therefore, in this density region, the process is mainly governed by stress relaxation. On the other hand, when the runs are started from $\rho$ = 1.53, a rapid increase of the pressure is observed at the beginning of the runs and fluctuation seems to continue during the healing process. In this plot, the pressure at 300 K becomes ~0 in the beginning and therefore clear temperature dependence is not observed for the pressure, in spite of the fact that the changes in density in these regions show temperature dependence. These results mean that the longer time process is affected by the thermally activated process such as jump motions of ions as already mentioned for disilicate systems. The healing process proceeds by the fluctuation of the pressure but not necessarily caused by negative pressures in longer time regions. Since the pressure of the system tends to become ~0 at the beginning of the run as shown in Fig. 7.16, the healing discussed here is not just shrunk by the pressure.

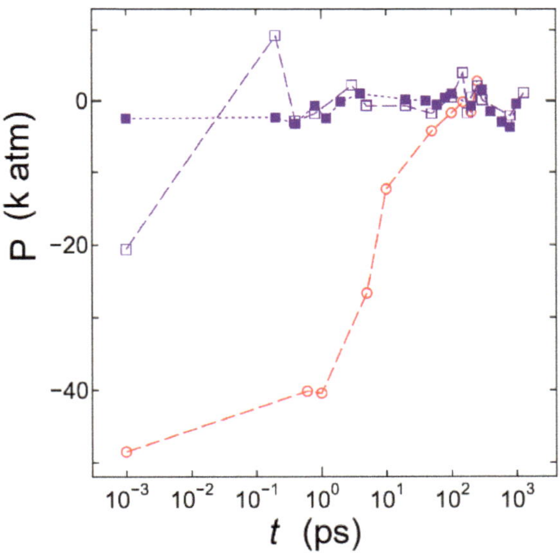


**Figure 7.16** Changes of pressures during the self-healing processes observed in porous lithium metasilicate systems during *NPT* runs at 600 K (open marks), where the pressure of the system is plotted against a logarithmic scale of time. For the lower density system, a result at 300 K (filled square) is also shown. The same marks and colors as in Fig. 7.14 are used here. In the system started from the higher density, negative pressure was observed during the process, while the system started from the lower density, fluctuation of pressure was observed during the process.

# 7.9 Perspective of Research for Healing Processes

## 7.9.1 Applications of Healing Properties

Applications of the healing process are emerging in many fields. Here we discuss the possible direction of the healing mechanism related to the porous lithium silicate systems. Recently, lithium disilicate and related materials have been used as dental ceramics. Due to its hardness, stability, and high thermal shock resistance, it is expected to play roles as the dental materials [28]. For example, IPS Empress® glass-ceramics (Ivoclar Vivadent) consists of 70% of needle-like lithium disilicate crystals embedded in a glassy matrix [29] and appear to have a good quality to fulfill the dental standard for abrasion behavior, chemical durability, and optical properties such as translucency of all glass-ceramics. As shown in the previous section, self-healing property of Li rich composition such as in lithium metasilicate can have a better self-healing ability. In this kind of application, a self-healing ability is one of the desirable properties, although it may have a trade-off relation with some of the other ones such as hardness.

## 7.9.2 Multifractal Nature and Possible Scale Gaps in the Problems of Healing and Nanofractures

Molecular dynamics simulations in these systems and related ones are useful for examining the mechanisms self-healing as well as mechanical properties such as nanoductility, nanofractures in nanostructured materials. For a measure of the roughness, multifractal analyses are often used after the pioneer works by Mandelbrot and collaborators [30]. For example, Vernède and collaborators [31] have examined a surface roughness of three systems, that is, an aluminum alloy, a mortar, and a sintered glass beads ceramic. The roughness of the fractured specimen is measured using a mechanical stylus profilometer. They have introduced the quantity $(\mathbf{x})$:

$$\omega_{\varepsilon}(\mathbf{x}) = \frac{1}{2}\log(<\delta h(\mathbf{x},\delta\mathbf{x})^2>) - \Omega\varepsilon, \tag{7.9}$$

where $\varepsilon = |\delta\mathbf{x}|$ and $\delta h(\mathbf{x}, \delta\mathbf{x}| = h(\mathbf{x} + \delta\mathbf{x}) - h(\mathbf{x})$ is the local slope of the surface in the direction of $\delta\mathbf{x}$. The value $\Omega$ is chosen so that the average of $\omega_\varepsilon(\mathbf{x})$ over all $\mathbf{x}$ becomes 0. In Fig. 7.17, the distribution of height fluctuations $P(\delta h|\delta r)$ at various scales $\delta r$ for the three samples considered is shown. The distribution found in Fig. 7.17 having a tail part and a truncation seems to be comparable to the Lévy distribution discussed in Chapter 5. In Fig. 7.18, spatial correlations represented for $\omega_\varepsilon$ computed at different scales $\varepsilon$ for three materials are shown. The cut-off length $\zeta$ is represented for each case. They have reported that the length scale $\zeta$ can separate the two regions of the surface as shown in this figure. Above this value, slope amplitudes are uncorrelated, and the fracture surface is mono-affine, while below $\zeta$, long-range spatial correlations lead to a multifractal behavior of the surface, which is reminiscent of turbulent flows. The size of $\xi$ was reported to be 50 ± 9 μm for ceramic in that work.

As found in this example, nanoscopic mechanical properties are not necessarily directly connected to the macroscopic behavior of the system.

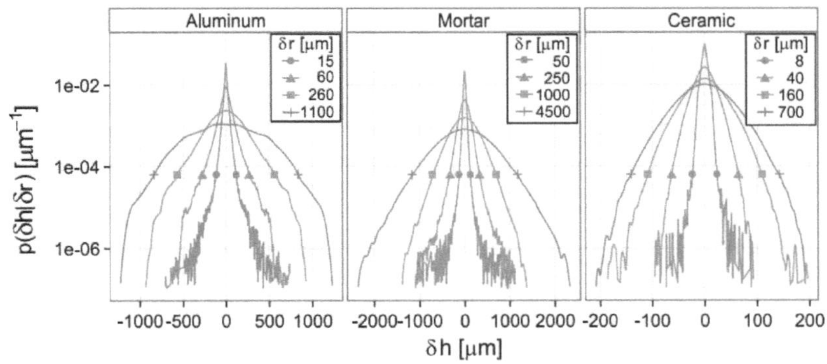


**Figure 7.17** Distribution of height fluctuations $P(\delta h|\delta r)$ at various scales $\delta r$ for the three samples considered. Reprinted with permission from Vernède, S., Ponson, L., Bouchaud, J.-P. (2015), Turbulent fracture surfaces: A footprint of damage percolation? *Phys. Rev. Lett.*, 114, 215501 (1–5). Copyright (2015) by the American Physical Society.

Even though multifractality is commonly found in experiments and MD simulations, one needs to be careful to connect the result of MD simulations of nanostructured systems to the macroscopic realistic systems directly. Time and length scale regions examined are not necessarily comparable. Furthermore, mechanical properties of the system related to the healing processes might depend on factors not only roughness of the surface but also defects, mixing of ordered and disordered phases in experimental conditions. Similar problems will be found for the study of fractures. The length scale of μm order mentioned above cannot be immediately applicable to the nanostructured materials. Therefore, it seems to be necessary to examine the problem of the scale-gaps and to clarify such characteristic length scales in nanostructured materials.

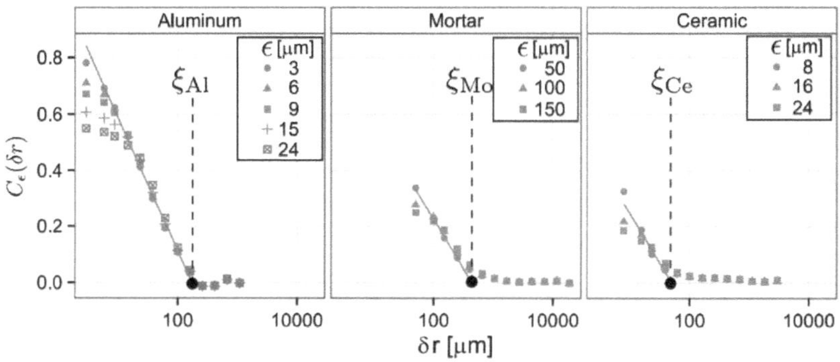


**Figure 7.18** Spatial correlations of $\omega\varepsilon$ for the three materials considered. The correlations are represented for $\omega\varepsilon$ computed at different scales $\varepsilon$. The cut-off length $\zeta$ is represented for each case. Reprinted with permission from Vernède, S., Ponson, L., Bouchaud, J.-P. (2015). Turbulent fracture surfaces: A footprint of damage percolation? *Phys. Rev. Lett.*, 114, 215501(1–5). Copyright (2015) by the American Physical Society.

Further systematic research for the relation between the self-healing ability and other mechanical properties is required. A comparison of details with experiments including size-dependent behaviors will be also useful. Again, multifractal analyses can be a common tool, which can give a connection between the structures and processes including the power law behaviors found in healing processes.

There is a possibility that through such analyses, healing processes and fractures can be understandable on the same basis, although one may also find some differences between them.

## 7.10 Conclusion

Two topics have been discussed in this chapter. One is concerned with the heterogeneous dynamics in porous systems in *NVE* conditions and the other is concerned with the formation of larger voids and a self-healing process of the porous systems in the subsequent runs in the *NPT* conditions.

In Chapter 6, enhancement of the dynamics of Li ions with the maximum is shown and the diffusive motion of Li ions is found to start from an early time (NCL) region of MSD corresponding to caged ion dynamics. The results demonstrate that the loosening of the cage in porous glass contributes to the enhancement.

As shown in the first part of this chapter, enhanced dynamics in porous lithium disilicate are heterogeneous dynamics with a non-Gaussian character as clearly proved by MD simulations. Changes in the characteristic times are more closely examined as the function of the density (porosity) and it was corroborated that the maximum of the inverse of the characteristic times is already found in the NCL region. From both changes of the characteristic time of MSD and changes in the power law exponent, it is shown that the changes in the motion in a longer length scale also contribute to the enhancement of the dynamics in porous systems, differently. In conclusion, dynamic heterogeneity of the system is affected by introducing pores in a complex manner in porous systems. Since the systems are sensitive to modifications, porous materials can be a useful platform to examine the complex behaviors of heterogeneous dynamics and the mechanism of glass transition related to such behaviors.

In the second part of this chapter, mechanisms of the self-organization of large voids in *NVE* runs and the self-healing process in *NPT* runs have been also examined. There are several time regions in healing processes and differences in these processes are discussed. In the healing process, it is found that the stress relaxation of the system plays an important role. The perspective of MD simulations in this field has also been discussed.

# Appendix

## 7.A.1 Summary of MD Simulations and Modeling of Porous Silicates

The lithium silicate systems examined here have been explained in the appendix in Chapter 6. Here we briefly summarize them for the reader of this chapter. The system examined is lithium disilicate, which is located in the silica-rich region of $Li_2O$–$SiO_2$ systems, and pores are introduced by the scaling of volume and positions. Each original and porous lithium disilicate ($Li_2Si_2O_5$) system examined contains 3456 atoms in the basic MD box with a periodic boundary condition. The following model function was used:

$$\phi_{ij} = \frac{z_i z_j e^2}{r} + f_0(b_i + b_j)\exp\left(\frac{a_i + a_j - r}{b_i + b_j}\right) - c_i c_j r^{-6} \tag{7.A.1}$$

The model potential consists of the Coulombic term, the pair potential function of the Gilbert–Ida type [32, 33] with the $r^{-6}$ for correction of the softness of the oxygen atom. The value $r$ is the distance between atoms and $a_i$ is the effective radius and $b_i$ is the softness parameters of the atom $i$ with a constant $f_0$ (=1 kcal Å$^{-1}$ mol$^{-1}$ = 4.184 kJ Å$^{-1}$ mol$^{-1}$). Potential parameters were derived based on the *ab initio* molecular orbital calculations [3, 34]. The Ewald method was used for the calculation of the Coulombic force. A cut-off distance was chosen to be 12 Å for the calculation of both the repulsive force and that for the real space term of the Coulombic force. The procedure for preparing porous systems is as follows. First, lithium disilicate in the glassy state was obtained by the rapid cooling (~1 K ps$^{-1}$) from the melt at 3000 K with the combination of constant pressure condition and temperature scaling to the target temperatures (2000, 1700, 1400, 1200, 1000, 800, 700, and 600 K). Systems were quasi-equilibrated at each temperature in the *NPT* condition and then following runs in the *NVE* condition were done. Then porous systems were prepared by the scaling of the volume and position of particles at 600 K. Similar methods had been previously used to prepare the porous silica [35, 36]. After the expansion of the system, each system was equilibrated during 300,000 steps run in the *NVT* condition and the following *NVE* runs are used for analyses. The time step used in the MD run was 1 fs for almost of porous systems. The densities $\rho$ examined are 2.47 (original), 2.30, 2.13, 1.98, 1.84 and 1.58. As the porosity, $P_0$, was found to be proportional to $-\rho$, the results are shown as a function of density.

That is, $P_0 = -17.57(\rho) + 100.08$, where $P_0$ is given in percent and density is in g $cm^{-3}$.

Simulations of porous metasilicate are also done to examine the enhanced diffusion and the self-healing process. Some results of the latter are shown in [27] and this chapter and details will be shown elsewhere. Some results of the self-healing process in porous lithium disilicate and metasilicate have been reported in several meetings [37, 38].

Recently, Zhang and coworkers pointed out that our potential model for sodium silicate is rather ductile in the *NPT* condition, while the behavior in the *NVT* condition is reasonable [39].

They concluded that one should obtain the stress-strain behavior in the *NPT* conditions and therefore our potential model for sodium silicate is not suitable for studying fractures.

Of course, in general, potential parameters are not necessarily optimized to study fractures of systems and therefore, some cautions are required for research of them. However, here we note that the nano-ductility and nano-fracture discussed in MD in certain conditions are not necessarily directly comparable to the real experiments under the atmospheric condition. These processes also show time-dependent properties. Further study for the condition of "determination of the stress-strain curve under the constant pressure" will be required before concluding the validity of the potential parameters. This is because in such a case, the behavior of the system is determined by the competition between the stress relaxation to form fractured structure and the self-healing process under the fluctuating pressure, where both processes have different time constants.

## References

1. Habasaki, J., Ngai, K. L. (2018), Heterogeneous dynamics in nano porous materials examined by molecular dynamics simulations

effects of modification of caged ion dynamics, *J. Non-Cryst. Solids*, 498, pp. 364–371. https://doi.org/10.1016/j.jnoncrysol.2018.03.026.

2. Habasaki, J. (2016), Molecular dynamics study of nano-porous materials—Enhancement of mobility of Li ions in lithium disilicate in *NVE* conditions, *J. Chem. Phys.*, 145, p. 204503.
3. Habasaki, J., León, C., Ngai, K. L. (2017) *Dynamics of Glassy, Crystalline and Liquid Ionic Conductors: Experiments, Theories, Simulations*, 1st ed. (Springer International, Switzerland) and references therein.
4. Ngai, K. L. (2011), *Relaxation and Diffusion in Complex Systems*, Springer, New York.
5. Einstein, A. (1905), Über die von der molekularkinetischen Theorie der Wärme geforderte Bewegung von in ruhenden Flüssigkeiten suspendierten Teilchen, *Annalen der Physik*, 322(8), pp. 549–560.
6. Loeb, A. L. (1976, 1991), *Space Structures* (Addison-Wesley, 1976; Birkhäuser, Boston, 1991).
7. Habasaki, J., Ngai, K. L. (2015), Rigidity and soft percolation in the glass transition of an atomistic model of ionic liquid, 1-ethyl-3-methyl imidazolium nitrate, from molecular dynamics simulations—existence of infinite overlapping networks in a fragile ionic liquid, *J. Chem. Phys.*, 142, 164501 (1–19).
8. Habasaki, J. (1990), Molecular dynamics study of the glass formation in $Li_2SiO_3$ system, *Molec. Phys.*, 70, pp. 513–528.
9. Tripol'skii, A. I., Strizhak, P. E. (2007), The effect of fractal dimension of porous catalysts on the activation energy of co oxidation, *Theor. Exp. Chem.*, 43, pp. 50–53.
10. Habasaki, J., Okada, I., Hiwatari, Y. (1997), Fracton excitation and Lévy flight dynamics in silicate glasses, *Phys. Rev. B*, 55, pp. 6309–6315.
11. Rahman, A. (1964), Correlations in the motion of atoms in liquid argon, *Phys. Rev.*, 136, pp. A405–A411.
12. Habasaki, J., Ngai, K. L., Hiwatari, Y. (2002), Molecular dynamics study of cage decay, near constant loss, and crossover to cooperative ion hopping in lithium metasilicate, *Phys. Rev. E*, 66, 021205(1–11).
13. van Hove, L. (1954), Correlations in space and time and born approximation scattering in systems of interacting particles, *Phys. Rev.*, 95, pp. 249–262.
14. Klafter, J., Shlesinger, M. F., Zumofen, G. (1996), Beyond Brownian motion, *Phys. Today*, 49, p. 33, and references therein.
15. Nakao, H. (2000), Multi-scaling properties of truncated Lévy flights, *Phys. Lett., A*, 266, pp. 282–289.

16. Mantegna, R. N., Stanley, H. E. (1994), Stochastic process with ultraslow convergence to a Gaussian: The truncated Lévy flight, *Phys. Rev. Lett.*, 73, p. 2946.
17. Koponen, I. (1995), Analytic approach to the problem of convergence of truncated Lévy flights towards the Gaussian stochastic process, *Phys. Rev. E,* 52, pp. 1197–1199.
18. Mandelbrot, B. (1967), How long is the coast of Britain? statistical self-similarity and fractional dimension, *Science*, 156. pp. 637–638.
19. Ben-Avraham, D., Havlin, S. (2000), *Diffusion and Reactions in Fractals and Disordered Systems*, Cambridge University Press.
20. Habasaki, J., Ngai, K. L. (2008), Molecular dynamics study of the dynamics near the glass transition in ionic liquids, *Anal. Sci.*, 24, pp. 1321–1327.
21. Habasaki, J., Ngai, K. L. (2018), Heterogeneous Dynamics in nano porous materials examined by molecular dynamics simulations effects of modification of caged ion dynamics, *J. Non-Cryst. Solids*, 498, pp. 364–371. https://doi.org/10.1016/j.jnoncrysol.2018.03.026.
22. Sing, K. S. W., Everett, D. H., Haul, R. A. W., Moscou, L., Pierotti, R. A., Rouquerol, J., Siemieniewska, T. (1985). Reporting physisorption data for gas solid systems with special reference to the determination of surface-area and porosity (recommendations 1984), *Pure Appl. Chem.*, 57, pp. 603–619.
23. Clausius, R. J. E. (1870). On a mechanical theorem applicable to heat, *Philos. Mag. Ser., 4*, 40, pp. 122–127.
24. Wang, B., Yu, Y., Wang, M., Mauro, J. C., Bauchy, M. (2016), Nanoductility in silicate glasses is driven by topological heterogeneity, *Phys. Rev. B,* 93, p. 064202.
25. Paris, P., Erdogan, F. A. (1963), Critical analysis of crack propagation laws, *J. Basic Eng. Trans. ASME*, 85, pp. 528–534.
26. Hooker, J. N., Laubach, S. E., Marrett, R. (2014), A universal power-law scaling exponent for fracture apertures in sandstones, *GSA Bull.*, 126, pp. 1340–1362.
27. Habasaki, J., A molecular dynamics study of enhanced dynamics and self-healing processes in nano-porous lithium metasilicate systems, submitted to *Int. J. App. Glass. Sci.*
28. Denry I. L. (1996) Recent advances in ceramics for dentistry, *Crit. Rev. Oral. Biol. Med.*, 7, pp. 134–143.
29. Höland, W., Schweiger, M., Frank, M., Rheinberger, V. (2000), A comparison of the microstructure and properties of the IPS

Empress 2 and the IPS Empress glass-ceramics, *J. Biomed. Mater. Res.*, 53, pp. 297–303.

30. Mandelbrot, B., Passojat, D. E., Paullayt, A. J. (1984), Fractal character of fracture surfaces of metals, *Nature*, 308, pp. 721–722.
31. Vernède, S., Ponson, L., Bouchaud, J.-P. (2015). Turbulent fracture surfaces: A footprint of damage percolation? *Phys. Rev. Lett.*, 114, 215501 (1–5).
32. Gilbert, T. L. (1968), Soft-sphere model for closed-shell atoms and ions, *J. Chem. Phys.*, 49, pp. 2640–2641. https://doi.org/10.1063/1.1670463.
33. Ida, Y. (1976), Interionic repulsive force and compressibility of ions, *Phys. Earth Planet. Interiors*, 13, pp. 97–104.
34. Habasaki, J., Okada, I. (1992), Molecular dynamics simulation of alkali silicates based on the quantum mechanical potential surfaces, *Mol. Sim.*, 9, pp. 319–326.
35. Kieffer, J., Angell, C. A. (1988), Generation of fractal structures by negative pressure rupturing of $SiO_2$ glass, *J. Non-Cryst. Solids*, 106(1–3), pp. 336–342.
36. Nakano, A., Bi, L., Kalia, R. K., Vashishta, P. (1994), Molecular-dynamics study of the structural correlation of porous silica with use of a parallel computer, *Phys. Rev. B*, 49, pp. 9441–9452.
37. Habasaki, J. (2018), Molecular Dynamics Study of Enhanced and Heterogeneous Dynamics in Porous Lithium Disilicates, ICG Annual Meeting 2018, September 23–26, Yokohama (invited).
38. Habasaki, J. (2018), Self-organization and self-healing processes in porous silicates examined by molecular dynamics simulations, Nanotechnology and material science congress (nanotech–2018), Nov. 12–13, Kuala Lumpur (Keynote).
39. Zhang, Z., Ispas, S. and Kob, W. The critical role of the interaction potential and simulation protocol for the structural and mechanical properties of sodosilicate glasses, October 2019, https://arxiv.org/pdf/1910.06181.

Chapter 8

# Full Atomistic Simulations of Nanocolloidal Solutions: Formation of Clusters, Aggregates, and Gels

In colloidal chemistry, attention is often paid to the interactions among colloidal units regardless of the details of structures of solvents and colloids. One of the popular approaches is based on the mean-field type approximation, in which the solution is treated as a homogeneous medium. However, the results of the recent MD simulations in atomistic-level have revealed that structural and dynamical changes of both solvent and salts play a role in spontaneous structure formations of aggregates and gels [1]. This situation will be discussed in this chapter.

## 8.1 Spontaneous Formation of Aggregates and Gels from the Nanocolloidal Silica in NaCl Solutions

In Ref. [1], coagulation of nanocolloidal silica in NaCl solutions is described based on molecular dynamics simulations using a

*Molecular Dynamics of Nanostructures and Nanoionics: Simulations in Complex Systems*
Junko Habasaki

ISBN 978-981-4800-77-8 (Hardcover), 978-1-003-04490-1 (eBook)
www.jennystanford.com

fully atomistic model. This approach enables us to examine the formation of clusters, aggregates, and gels without any assumption of mechanisms of formation and therefore new insights in colloidal chemistry can be brought.

For examining a colloidal system, a coarse-grained model is often used, and details of the solvent tend to be neglected, as already mentioned. A main reason for such treatments seems to be a large size difference between colloidal units and solvent molecules, which makes it difficult to treat the system in fully atomistic levels. Furthermore, dynamics during the formation of gels are slow, as observed in the glass transition of glass-forming materials, where heterogeneous dynamics are often found. In the case of the treatment of heterogeneous systems having several length scales, a large system and a long-time simulation are required. Therefore, molecular dynamics simulations of the colloidal system in fully atomistic models are challenging, although obviously, they are useful for a better understanding of coagulation processes in nanoscopic levels.

The effect of salt on the coagulation process is experimentally well known and tends to be explained by the DLVO (B. Derjaguin, L. Landau, E. Verwey, and T. Overbeek) model [2, 3] in many cases; however, many problems remain and are a subject of debate.

For example, the effect of electrolytes on dispersions depends on the kind of ions. Such a specific ion effect [4] is not fully explained by the theory. Furthermore, multivalent salt is known to be more efficient in reducing the effective charge of the colloidal solution. This multivalent ion effect [5, 6] is another example not to be well explained by the DLVO model.

In the present chapter, the origins of forces, the role of the solvent affected by salt for coagulation of nanocolloidal-silica—water–NaCl system will be discussed based on structural and dynamical changes.

Especially, the following changes in water structure cannot be clarified without atomistic and/or molecular level simulations. With the increased concentration of the NaCl, a pair correlation function of the OW–OW (OW means an oxygen atom in water) pair shows a sub-peak characteristic to the formation of high-density water (HDL) related to the so-called polyamorphism [7] of water.

Parallel with the structural changes of water, the network structure of the NaCl part also changes and these structures are affected by the presence of the silica part in the system as shown later. Therefore, the picture based on the interaction among colloidal units with a mean field approximation of solvent seems to need revision for considering coagulation processes.

## 8.2 Molecular Dynamics Simulation of Colloidal Systems

In the atomistic simulations shown in Chapters 8 and 9, potential parameters used in [1] are chosen to maintain consistency of quality and combinations as much as possible, although further improvement of these parameters might be necessary, depending on the purpose. For the colloidal silica and resultant silica part of aggregations and gels, the potential model developed by Tsuneyuki et al. [8] (the TTAM model) was used. It is worth to mention that the model of silica was not necessarily designed for the simulation of colloids and/or gels. However, it allows the breaks and restructures of the Si–O bond because it is determined based on the *ab initio* MO calculation with the deformation of an $SiO_4^{4-}$ + $4e^+$ cluster including changes in an Si–O distance. For water, an SPC/E [9] model is used, and it is treated using the shake algorithm with a constraint. This model water has been well tested (for example, see [10]) and parameters for interactions with NaCl salt are available [11]. The interaction between silica and $Na^+$ or $Cl^-$ was obtained from the CLAYFF Force Field [12].

Further details of the potential functions and parameters used are described in the Appendix and Ref. [1].

Ensembles used in each run will be shown in the next subsection. The DL_POLY (version 4.01) program [13] was used for the MD simulations.

### 8.2.1 Modeling of Colloidal Systems

One of the problems is how to prepare the complex structures formed by several components and or units used in MD simulations. In this work, equilibrated colloidal units surrounded by solvent molecules are prepared and then they are combined to form larger systems. This procedure will be described below.

### 8.2.2 Preparation of Colloidal Units

First, an initial colloidal unit composed of $SiO_2$ was prepared as an isolated partial system.

There are six atomic species in the system (Si, O, Na, Cl, OW, HW). If we would like to distinguish oxygen and hydrogen in water from those belonging to other species, notations OW and HW are used, respectively. Among combinations of pairs, some repulsive interactions can be excluded from further treatments when direct contact is rare.

Here we consider the colloidal units formed by silica networks.

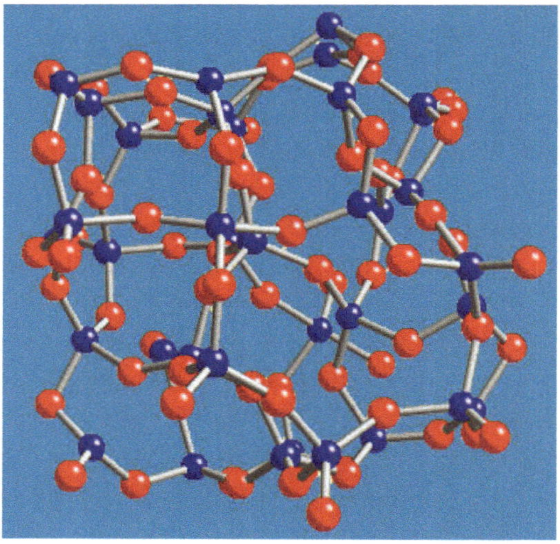

**Figure 8.1** An example of an initial structure of a colloidal unit represented using a ball-and-stick model.

First, a colloidal silica unit consisting of 108 atoms was equilibrated in the *NVE* ensemble with a periodic boundary condition (PBC), where the volume was fixed to the value obtained from the NPT ensemble (using a Nosé–Hoover thermostat and a barostat [14–16]) under atmospheric pressure. After equilibration at 300 K under constant pressure (0.1 MPa), the PBC was removed.

In the system with PBC, the Coulombic force is calculated by the particle-mesh method [17], while in the system without PBC, the Coulombic force was calculated as a direct sum. During this procedure, the temperature increased to allow the melting of the system and a cluster with a rounded shape was gradually formed. In Fig. 8.1, an example of a structure of a colloidal unit is illustrated using a ball-and-stick model. The characterization of the colloidal unit thus prepared can be found in Ref. [1].

The colloidal unit was solvated with SPC/E water within a cubic basic box. Then $Na^+$ and $Cl^-$ ions were added at random positions.

### 8.2.3 Preparations of the Complex System with Colloidal Units, Water and Salt

A solvated system containing one colloidal unit and salt was equilibrated at 300 K in the *NPT* condition with PBC. After equilibration, an isotropic basic cell was prepared for the large-scale MD simulations by creating 64 (4×4×4) repetitions of the solvated colloidal unit cell.

Six systems shown in Table 8.1 with PBCs were examined in Ref. [1]. Some descriptions of the systems, including the number of colloidal units and atoms, the number of $Na^+$-$Cl^-$ pairs, and the concentration of salt in water, with the status of the system, are provided in this table. These results and some new results will be discussed in this chapter. (Here the concentration is given as the ratio to water and not to solution. This choice seems to be better for comparison among systems because the status of after the system also depends on the amount of colloidal silica).

For each system, the MD simulation was first performed in the *NPT* ensemble, and the run in the *NVE* ensemble then followed the equilibration of the system.

The I-A, I-B, and I-C and II-A, II-B, and II-C systems examined are characterized as follows. In series I, with smaller water content, we have observed the systematic formation of clusters, aggregates, and a gel. In series II with larger water content, each system shows no large aggregates, including the reference system II-A without salt.

**Table 8.1** Characteristics and status of systems examined in Ref. [1], Chapters 8 and 9

| System | Number of colloidal silica units | Number of $Na^+$-$Cl^-$ Pairs | Number of water molecules | Concentration of NaCl in water (mol%) | Total number of atoms | Temperature (K) | Status |
|---|---|---|---|---|---|---|---|
| I-A | 64 (6912 atoms) | 128 | 8192 | 1.56 | 31744 | 300 | Aggregate |
| I-B | 64 | 320 | 8192 | 3.9 | 32128 | 300 | Aggregate |
| I-C | 64 | 320 | 8192 | 3.9 | 32128 | 400 | Aggregate, Gel |
| II-A | 64 | 0 | 15872 | 0 | 54528 | 300 | Clusters |
| II-B | 64 | 160 | 15872 | 1.01 | 54848 | 300 | Clusters and Aggregate |
| II-C | 64 | 320 | 15872 | 2.02 | 55168 | 300 | Aggregate |

Each system except I-C was equilibrated at 400 K until the energy of the system became nearly constant, and the temperature was then decreased to 300 K. The size of the basic cell for the MD simulations, $L$, of the complex system addressed the present work was 65–90 Å. To examine $g(r)$ at a long length scale, an additional run was performed in each system with a long cut-off length of $\sim L/2$ (for both the repulsive force in the real space and the real part of the electrostatic force), whereas a cut-off length of 12 Å was used in all other cases. Each time step was set to 1 fs. After the construction of the initial configurations, the systems were equilibrated for 1–3 ns. When the energies and structures of one 100,000-step run (1 step=1 fs) were essentially maintained in the subsequent 100,000-step run, the system was regarded as sufficiently equilibrated one, and the simulations were continued on a longer time scale. After the equilibration of the system, the ensemble was changed to *NVE*, using the volume obtained from the *NPT* run. The results for the time range of 400 ps to 2 ns (400,000–2,000,000 steps) at 300 K after the equilibration run were used in analyses.

The formation of a gel was observed in the following condition in I-C system. Beginning from the I-B configuration, the temperature was again increased to 400 K to accelerate coagulation. We observed the developments of infinitive networks in this system, in which the atoms became immobile as shown by the mean-squared displacement of the system (see Section 8.6). This system is regarded as a gel in the present model. The gel thus obtained was quasi-equilibrated one because of the slow dynamics.

### 8.2.4 Structural Changes with Coagulation Caused by Salt

The structural characteristics of each system in Table 8.1 will be discussed further in this subsection.

As naturally expected, the structural changes by the development of aggregates of colloidal silica were observed with increasing content of salt.

In the case of examining the structural changes on the colloidal surface or reaction with the coagulation process itself, the works by *ab initio* MD or classical MD using a reactive force field can be a useful and better approach; however, our concern here is equilibrated systems near the potential minimum after the coagulation processes. In such cases, potential parameters of silica, which allows the rearrangement of Si–O bonds, are good enough, even if they were not designed to represent the reaction process or chemical states on the surface. Both bonds breaking and reconnection allow structures after the long relaxation time to be meaningful. In fact, changes of the (quasi)-equilibrated structures are systematic as shown in the following subsections.

In MD simulations, initial configurations are not necessarily equilibrated ones, so that time development of the structures depends on the choice of them. Therefore, for a kinetic argument for the formation of aggregations or gels, different approaches might be required.

#### 8.2.4.1 Changes in network structures of silica and water parts

To learn the characteristics of networks, structures of the silica part were visualized, and then their fractal dimensions were examined. The calculated dimension for silica part was comparable to the experimental observations for the aggregation of silica colloids, and at longer length scales, super-aggregation was evident in the gelation process.

Clear changes in both the structure and mobility of the water were observed as the NaCl concentration increased, suggesting the importance of the solvent structures in these processes. The structural change in the water that was observed in the pair correlation function is interpreted as the change from a lower-density liquid to a high-density liquid (HDL) as shown later.

## 8.3 Changes in the Network Structures and the Formation of Gels

In both I-A and I-B, the colloidal units form one large aggregate. In the I-B system, which has a larger NaCl content, the connections among the units are more developed.

The instantaneous structures of the systems I-A and I-B obtained from the MD simulations after equilibration are depicted in Ref. [1]. Figures 8.2 and 8.3 show the structures of system II-A and system I-C after gelation, respectively. The former system is a reference without salt. A three-dimensional view of II-A reveals that some small clusters are separated from the largest cluster. The size of these clusters is several times larger than that of the original $SiO_2$ colloidal units in our systems.

As shown in this figure, spherical shapes of colloidal units are not completely lost in clusters and some $SiO_4$ units are connected by sharing edges (or corners, faces) to form a larger cluster. Naturally, in systems II-A and B, which contain a larger amount of water (by a factor of ~two) than systems I-A and I-B, the network is less developed. In the system I-C, following a short run (~several hundred ps at 400 K) after the long run at 300 K, a gradual shrinkage (approximately 19% in volume) of the system in *NPT* conditions is found. As shown later, this process occurred in parallel with the slowing down of the system including the water as well as the formation of three-dimensional (infinitive) networks of silica. The networks are formed of -Si–O–Si- bonds and the water molecules were trapped in the network. We regard this formation of a 3D percolated structure as gelation. In Fig. 8.3, the structure of the gel thus spontaneously formed is represented.

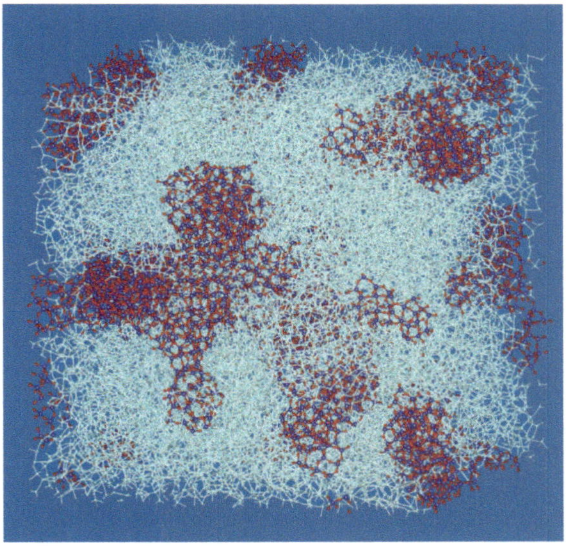

**Figure 8.2** Structure of system II-A at 300 K within a basic cell of MD. It contains 19 clusters in the basic cell of the simulation Blue: Si and Red: O are connected by bonds. Pale blue polyhedra: Ow (oxygen in water molecules) surrounded by Hw (Hydrogen in water molecules) atoms. Silica nanocolloids are separated by networks of water in this system.

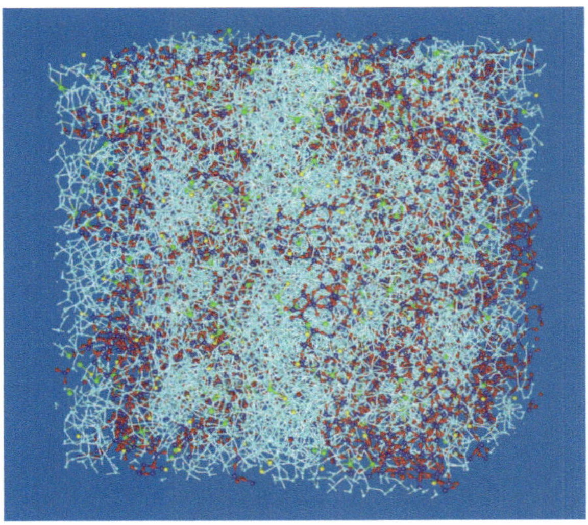

**Figure 8.3** Structure of the system I-C (after gelation). Green: Na, Yellow: Cl. The system forms infinitive networks of both silica and water networks with periodic boundary conditions.

The system seems to be more ordered than in the case of typical aggregates, suggesting that the long-range structure of the system may be affected by a periodic boundary condition. This finding suggests the existence of long-range cooperative motions during the formation of gels. This process may also be affected by the initial arrangements of colloidal units. The dynamics of each constituent of the system, including the solvent, changed considerably with the formation of the gel, as demonstrated later.

### 8.3.1 Reconstruction of Si–O Bonds during Formation of Gels and Distribution of $Q_n$ Structures

How does the number of connections in the network change with the formation of gels? The coordination number for Si around Si represents the possible connections in the network through Si–O–Si. The number of Si–O bonds in each system was counted using a cut-off length of 2.1 Å. The mean number of bridging oxygens, $N_{b\text{-}O}$ (for Si-(O)-Si connections), in each system was determined. The number was found to be 3.7–3.9 for these systems. From these results, one can conclude that the formation of gels occurred with the combination of bond-break and bond-formation with keeping the total number of bonds. Using the same definition of Si–O bonds, the distribution of $Q_n$ structures (where $n$ is the number of bridging oxygen atoms in the $SiO_4$ units) was determined to characterize each network system (see Fig. 8.4 for typical $Q_n$ structures). The distributions of $Q_n$ structures obtained by MD simulations are shown in Fig. 8.5. All systems show the peaks at $n = 4$ of $Q_n$ structures. In several systems, a negligibly small number of $Q_5$ structures was found. In the series of I in Fig. 8.5a, there are small differences in the distributions, while for the series of II in Fig. 8.5b, the distributions are almost coincident with each other. The cluster structure is thus formed by the reconstruction of networks, that is, both bonds breaking within each colloidal unit and the formation of new bonds among units occurred.

Further discussion of changes in $Q_n$ distribution will be given in Chapter 9.

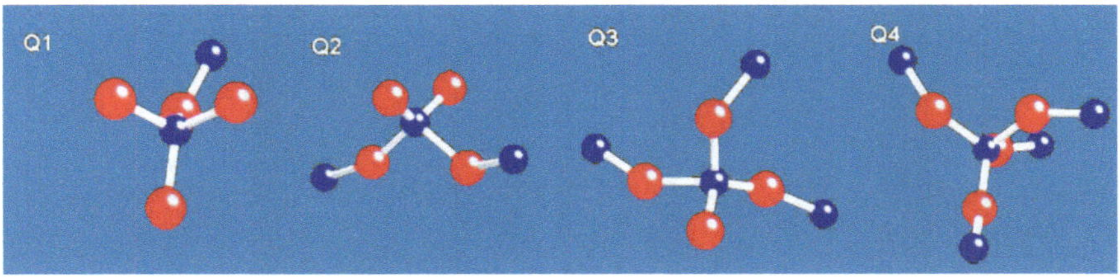


**Figure 8.4** Schematic description of $Q_n$ structures with $n$ = 1–4, where $n$ means a number of bridging oxygen atoms. Si atoms connected to the bridging oxygen atoms are also shown. Reproduced with some modification from Habasaki, J., Leon, C., Ngai, K. L. (20017), Chapter 9, Molecular dynamics simulations of silicate glasses, in *Dynamics of Glassy, Crystalline and Liquid Ionic Conductors*, 2017, Springer, with permission.

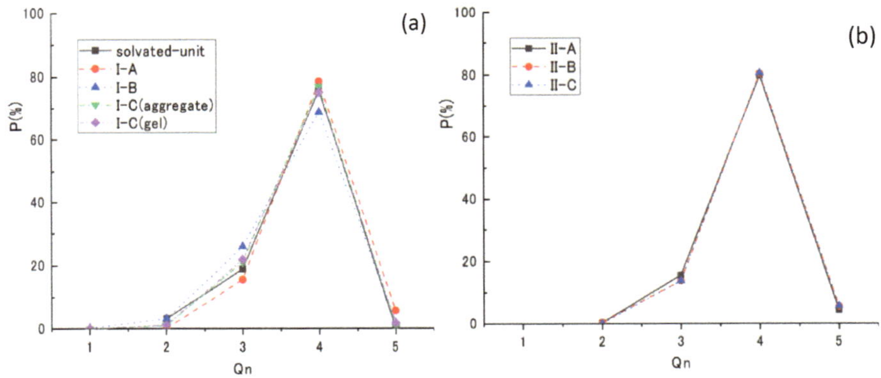


**Figure 8.5** Distribution of $Q_n$ values (number of $n$ is taken as axis number). (a) For colloidal silica unit surrounded by water and systems in series I with smaller water contents. (b) For series II with larger water contents. Three kinds of distributions overlap in this figure.

## 8.4 Pair Correlation Functions and Running Coordination Numbers

A pair correlation function, $g(r)$ and a running coordination number, $N(r)$, are useful to characterize the networks and the formation of gels. The pair correlation function, $g(r)$, for the Si–Si pairs (connected by bridging oxygen) was determined as shown in Fig. 8.6. In Fig. 8.6a, the pair correlation functions for the Si–Si pairs

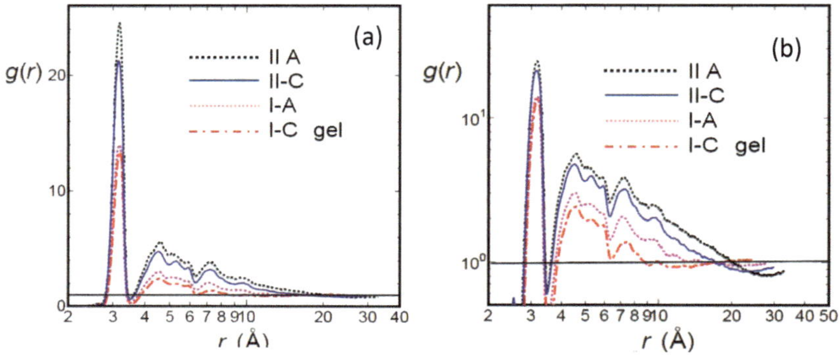


**Figure 8.6** (a) Pair correlation functions of Si-Si pairs in the systems I-A, I-C, II-A, and II-C on a log-log scale. A power-law region and a region corresponding to $g(r) \sim 1$ with some deviations are observed. The $g(r)$ function displays a clear minimum near $r$ = 30 (Å) in the system without salt, in which the clusters are isolated. Minimum of the function corresponds to the maximum of the potential of the mean force $W_{ij}$. That is, it corresponds to the barrier of the coagulation. (b) Pair correlation function of Si–Si pairs in double logarithm plots. The same pairs as in (a) are contained. The minimum in each curve is clearer. Figures are reproduced from Habasaki, J., Ishikawa, M. (2014), Molecular dynamics study of coagulation in silica-nanocolloid–water–NaCl systems based on the atomistic model, *Phys. Chem. Chem. Phys.*, 16, 24000–24017, with permission from the PCCP Owner Societies.

of the systems I-A, I-C, II-A and II-C are plotted on a semi-log scale, whereas in Fig. 8.6b, these correlation functions are plotted on a double logarithmic scale. The series II systems exhibit larger peaks than the series I systems, which indicate more isolated clusters. Remarkable features observed are systematic changes in peak heights and positions of minima in the functions. I-A and I-C exhibit smaller peak heights than II-A and II-C. For a higher concentration of NaCl, the Si–Si peak heights for the first and several subsequent shells (for dense regions of the networks) decrease, and these dense regions become smaller in both series of systems. A small minimum (deviation from $g(r)$ = 1) is observed in the long-length-scale regions ($r$ > 12~35 Å) of the dilute systems before $g(r)$ = 1 is attained. This trend can be more clearly observed when a logarithmic scale is used for the $g(r)$ value, as shown in Fig. 8.6b. The minimum is located at approximately half of the mean distance between clusters or aggregates when these structures are separated by the solvent.

The existence of this minimum is consistent with the results of off-lattice DCLA calculations performed by Hasmy et al. [18]. In the pair correlation function, the $r = 0$ position of the function is common to all Si atoms, including those located on the peripheries of clusters. The size of these clusters is several times larger than that of the original $SiO_2$ colloidal units in our systems. The largest deviation from 1 is observed in the system without salt (II-A) at approximately $r = 30$ Å. A decrease in the peak heights of $g(r)$ in the short-length-scale region can be observed with increasing NaCl concentration for I-B (not shown); hence, changes occur with coagulation, while, higher temperature of system I-C has no significant effect on the function. Thus, in system I-C, the power-law region below $r \sim 12$ Å represents the spreading of the network of clusters, and the longer-length-scale region represents the development of connections among these structures to form a gel.

## 8.5 Barrier for Coagulation

### 8.5.1 Concept of the Potential of Mean Force

The structure found in $g(r)$ of Si–Si pairs shown in Fig. 8.6 is also useful to understand the barrier of coagulation process, which is related to the maximum observed in the pair correlation function, $g(r)$ of Si–Si pairs.

As shown in the previous section, the minimum of $g(r)$ is located at approximately half of the mean distance between clusters or aggregates when these structures are separated by solvent. The existence of the minimum seems to be important to understand complex structures in gels and its mechanism of the formation.

Here we explain the physical meaning of the barrier based on the concept of the potential of mean force (PMF), which was introduced by Kirkwood [19]. Generally, the concept plays an important role in statistical mechanical theories of liquids. For equilibrated or nearly equilibrated systems, the PMF, $W_{ij}(r)$, is related to the pair correlation function $g_{ij}(r)$ as follows:

$$W_{ij}(r) = -k_B T \ln g_{ij}(r) \tag{8.1}$$

From this functional form, the minimum of $g(r)$ observed in Fig. 8.6 corresponds to the barrier for effective repulsive interaction among colloidal clusters. That is, in $g(r)$, it corresponds to the boundary region separating clusters. As expected, the effective repulsive force is the largest for the system without salt, for which the deepest minimum is observed. In the gel, the length of the power-law region is the shortest among all investigated systems, and the minimum is not so clear. In this case, the structure of the original colloidal units rapidly decayed and spread to form the gel. The strength of the effective repulsive interaction among aggregates is controlled by the salt concentration and the packing density of $SiO_2$ ($SiO_2$: water ratio) as systematic changes of $g(r)$ with concentrations.

In the case of the DLVO model and related ones, one assumes the potential consists of the repulsive and attractive parts. In the case of MD simulations, the origin of each force is clear from the potential functions used and the $g(r)$ is determined by the sum of forces. Analytically, the force acting for an $i$-th particle is represented by [20–22]

$$-\frac{\partial W}{\partial r}=\frac{k_B T}{g(r)}\bullet\frac{\partial g(r)}{\partial r}=\frac{\int\cdots\int\left(-\frac{\partial U_N}{\partial r_i}\right)\exp(-\beta U_N)dr_{n+1},\ldots,dr_N}{\exp(-\beta U_N)dr_{n+1},\ldots,dr_N}, \quad (8.2)$$

where the $U_N$ is for internal energy for $n$ body potentials of $N$ particles. In the case of the pair potential, $n = 2$.

For obtaining the PMF (or related free energy) from simulation runs along with reaction coordinates, techniques such as umbrella sampling [23, 24] are required for obtaining good statistics. For example, Allen et al. [25] have examined the potential of mean force to understand the behavior of particles in narrow channels. However, this is not the case for the discussion of the PMF from the pair correlation function, $g(r)$, using Eq. (8.1), because the pair correlation function can be obtained with good statistics in MD using many numbers of initial configurations and many particles. That is, the trend in the PMF can be determined based on the behavior of $g(r)$. From Fig. 8.6, we have judged that the system sizes are large enough to observe and argue the systematic changes in the minimum position of the $g(r)$

related to the separation of colloids or clusters. If the calculation resources and times were available and if actual calculation could be done within an acceptable real time, larger systems may be favored for obtaining precise characteristics of the barrier. Although the explanation was given for the Si-Si pair, it is needless to say that one must include interactions of all pairs, to consider the structure of the system.

### 8.5.2 Structure Surrounding Colloidal Silica–Running Coordination Numbers

In Fig. 8.7, the structure of a cluster taken from the gel, defined by a sphere with a radius of 12 Å around a certain Si atom, is shown. One can observe how structures are mixing when it is observed from a central Si ion. As shown in this figure, the silica part and water part are mixing as well as Na ions and Cl ions. In Fig. 8.8, pair correlation function, $g(r)$, of pairs of species in I-C and II-A are compared for the power law region. All pairs including Si atoms are shown here. From these curves, one can understand the mean behavior of how structures are mixing when we observe it from all central Si ions.

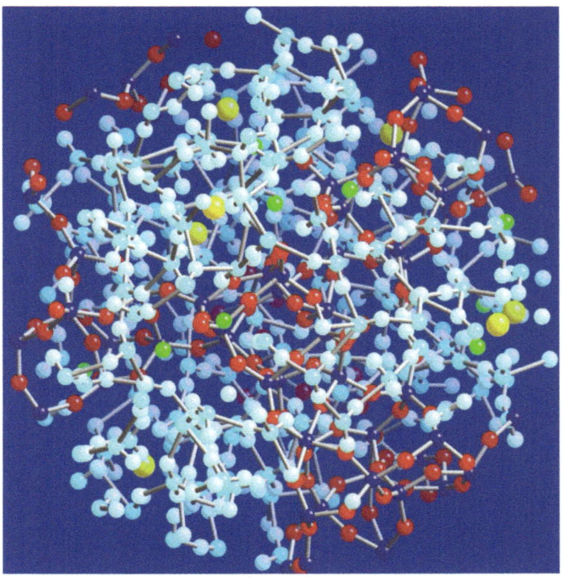

**Figure 8.7** A structure of a cluster (radius = 12 Å) taken from a gel. Blue: Si, Red: O. Green: Cl. Yellow: Na. Blue: OW. Pale blue: HW.

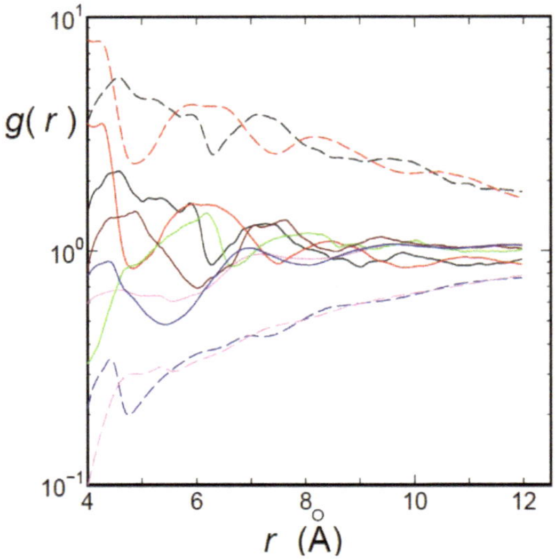


**Figure 8.8** Comparison of pair correlation functions, $g(r)$, in the power-law region for systems I-C (solid curves) and II-A (dashed curves). Black: Si–Si, Red: Si–O, Blue: Si–OW, Pink: Si–HW, Green: Si–$Na^+$, Brown: Si–$Cl^-$. In II-A, the contribution of the $SiO_2$ component decreases with increasing $r$, while that of the water component gradually increases. In I-C, the slopes of these components flatten, indicating the mixing of these structures. This figure is reproduced from Habasaki, J., Ishikawa, M. (2014), Molecular dynamics study of coagulation in silica-nanocolloid–water–NaCl systems based on the atomistic model, *Phys. Chem. Chem. Phys.*, 16, 24000–24017, with permission from PCCP Owner Societies.

In II-A, upper curves (silica part) for Si–Si and Si–O pairs are fairly separated from those for Si–OW and Si–HW pairs, while the curves are not separated in I-C (gel). Here OW is for the oxygen and HW is for the hydrogen of water molecules.

It is useful to compare the running coordination numbers $N(r)$ for every species around Si atoms to understand how $g(r)$ of Si–Si pairs is determined in the system.

The running coordination number $N(r)$ is defined as follows:

$$N_{\mathrm{coord}}(r) \equiv \rho_j \int_0^r 4\pi r^2 g(r)dr, \tag{8.3}$$

where $\rho_j$ is the number density of the species $j$ in the system.

In Fig. 8.9, corresponding running coordination numbers are shown for I-C and II-A systems. As seen in these figures, the numbers corresponding to HW and OW around Si atoms differ notably in these systems. The running coordination numbers of the water components rapidly increase with increasing *r*. That is, larger changes in the *N*(*r*) slopes are observed for Si–HW and Si–OW than for Si–Si or Si–O in both II-A and I-C. In system I-C, a larger amount of solvent exists near Si than in system II-A. This means that the structure of water is a counterpart of the silica network.

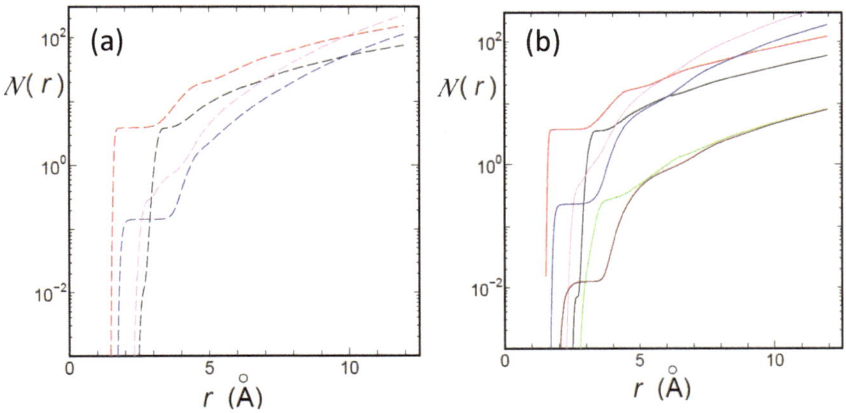


**Figure 8.9** (a) Running coordination numbers for System II-A. (b) Running coordination numbers for System I-C. The colors in (a) and (b) are the same as in Fig. 8.7. The functions for the Si–Si and Si–O pairs are the counterparts of the functions for the water components (Si–OW and Si–HW), which are affected by the salt concentration. This figure is reproduced from Habasaki, J., Ishikawa, M. (2014), Molecular dynamics study of coagulation in silica-nanocolloid–water–NaCl systems based on the atomistic model, *Phys. Chem. Chem. Phys.*, 16, 24000–24017, with permission from PCCP Owner Societies.

Of course, the contribution of the salt in I-C increases as a function of *r*; however, the coordination number of $Na^+$ and/or $Cl^-$ in System I-C (gel) is 20~30 times smaller than that of water in the shorter-*r* region (at around 3 Å) and ~10 times smaller in the longer-*r* region. This result means that the concentration of NaCl is not homogeneous when we consider the nanoscopic structures of colloids solution and/or the formation of gels. From these results, we can conclude that the system is not a simple mixture of silica part and solvent but each component of silica, water, and NaCl plays a role to form complicated nanostructures. Further details of the structures of the silica part such as fractal dimension and connections of $SiO_4$ units represented by a dendrogram will be shown in Chapter 9.

### 8.5.3 Comparison with Description by the DLVO Model

The DLVO model [2, 3] is often used to explain the behavior of the colloidal systems. This model accounts for two types of forces that are present in a stable colloid: the van der Waals force and electrostatic repulsion. The total interaction potential can be calculated as a function of distance, and the colloid is stable when the two forces balance each other.

It can successfully explain the stability of colloidal systems in several cases but not necessarily adequate for other cases.

One can find that the explanation of barrier for coagulation shown in the previous section is not necessarily the same as that based on the DLVO model, concerning with the origin of the force determining the barrier of the coagulations and each contribution of terms in the potential functions can be separated in MD simulations if necessary. Therefore, the information from the atomistic simulations will be useful for the origins of forces in related systems without assuming the DLVO model a priori.

## 8.6 The Structural and Dynamical Changes of Water with Variations in the NaCl Concentration

### 8.6.1 Structural Changes of Water and Polyamorphism

In this subsection, structure of the water part is examined. Both structures and dynamics of the water change with variations in the salt concentration. This means that the salt affects not only the interaction among colloidal units but also the interaction between the colloid and the water. The pair correlation functions, $g(r)$, for HW–HW, HW–OW and OW–OW pairs are presented in Figs. 8.10a,b for systems I-A and I-C after gelation, respectively.

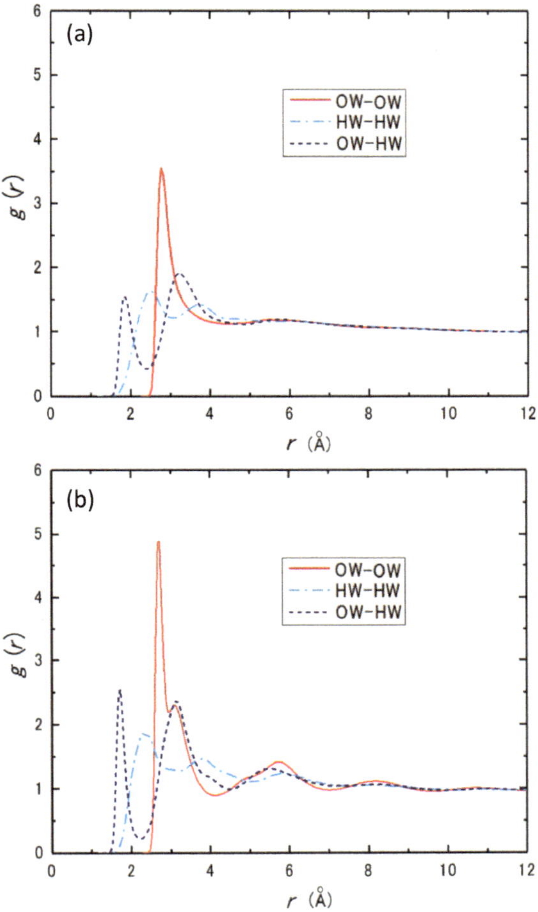


**Figure 8.10** Pair correlation functions of water for systems (a) I–A and (b) I–C (after gelation). With the addition of salt, the first peak in the $g(r)$ for the OW-OW pair splits, and this splitting becomes clearer in the gel. A long-range structure indicated by waves in the longer-$r$ region also appears. Some small changes are also observed for the other pairs. These changes are like those associated with the LDL-to-HDL transition in water (see text).

The comparison of these two figures shows several differences. First, the peak heights of functions in I-C are larger than I-A. The first peak of the OW–OW has a shoulder or a split of the peak into two, and the function has oscillations continuing for a longer length. The peak height of the pair correlation function corresponding to OW–HW pairs also increases with increasing NaCl concentration. These trends are consistent with the decrease in the mobility of the water. Thus, this sharpening seems to be caused by the trapping of water in the gel.

Is the splitting just by the slowing down of the motion? To answer this question, the coordination number was examined. The coordination number of OW around OW in the pure SPC/E water is 4.0 at the first minimum (at ~3.2 Å) and the location of the second peak is well separated from this feature. In the gel (I-C) observed in the present simulation, the coordination number corresponding to this structure is found to be 10.3 at the length just after the shoulder (~4.0 Å). Therefore, in the gel, the second shell collapsed and the clear minimum at around 3.2 Å is missing. Therefore, the second part of the splitting is not just by the arrest of the motion but with the rearrangement of the network structure of water.

For such structural changes, consideration of further shells is necessary.

### 8.6.2 Gradual Changes of Density of Water during the Formation of Gel

It is interesting to note that the structural change in the water that is observed with variations in the NaCl concentration is comparable to the transition from a low-density liquid (LDL) to a high-density liquid (HDL) that has been observed in several works [28–30] related to polyamorphism [31] of water. Of course, some caution is required that the water at room temperature is not typical LDL but is a mixture of the different types of coordination polyhedra for hydrogen bonding [32]. Soper et al. [30] have reported the observation of structural changes in water with varying pressure at 268 K through neutron diffraction. In their work, a transition from open, hydrogen-bonded tetrahedral structures to non-tetrahedral O–O–O angles and collapsed second coordination shells were observed with increasing pressure. The large volume reduction was observed during the gelation process (approximately 19%) in this work. More details with a historical view will be shown in Chapter 9.

### 8.6.3 Dynamical Changes in Water and Other Constituents

As usual, the equilibration of the system in the MD simulations was evaluated in terms of the energy of the system and the behavior of the structures. However, because gradual changes in the clusters can be observed with small changes in energy, it is not clear how long the time scale is required for the study of the clustering and gelation processes in the MD simulations. Probably, based on the mobility of each species, one can judge which local structures are equilibrated and which are not. Mean-squared displacement of species "*a*" can be defined as follows:

$$<r_a^2(t)> = \left\langle \{\sum_{i=1}^{N_a} (\boldsymbol{r}_{ai}(t) - \boldsymbol{r}_{ai}(0))^2\}/N_a \right\rangle, \tag{8.6}$$

where $\boldsymbol{r}$ is a positional vector, while $r$ is a displacement of particle $i$ of species "*a*" for the radial direction. Common MSD behaviors with several distinct stages are observed in many systems [33, 34] with slow dynamics. It is also shown in Chapter 5 for lithium silicates. As clarified for the motion of Li ions in lithium silicate glasses [35, 36], after a time $t_{\mathrm{dif}}$, the MSD indicates a diffusive regime. At that time, particles move from original positions to a certain mean neighboring distance.

In Fig. 8.11a, an MSD of each species in the quasi-equilibrated state for I-A is presented as a function of time. The MSDs of Si and O are small and less than 3 Å$^2$ within a 1 ns run. The MSD of $Na^+$ seems to exhibit an increase, parallel to the MSD of colloidal silica. The motion of $Cl^-$ is larger than that of $Na^+$, and the slope seems to be nearly comparable to that of water. Therefore, the motions of $Na^+$ and $Cl^-$ are essentially decoupled. For the system I-B, the silica component is immobile and the slopes of the MSDs of $Na^+$, $Cl^-$ and water decrease. The motions of $Na^+$ and $Cl^-$ are comparable with each other and seem to be coupled. The higher concentration of NaCl seems to make the water more rigid, because of the strong interaction among the salts. The role of $Cl^-$ is non-negligible in this case. For example, HW atoms of water molecules tend to gather near $Cl^-$ ions, and they affect the structure of the water. They also affect the distance between $Na^+$ and $Cl^-$.

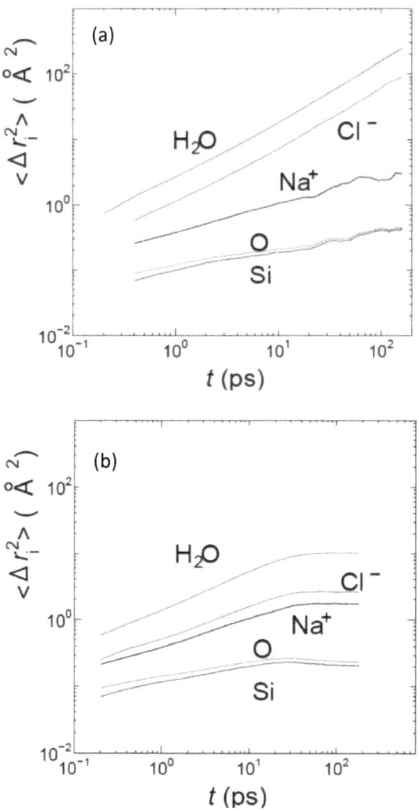


**Figure 8.11** (a) MSD of each species in system I–A. (b) MSD of each species in system I–C. With the formation of infinitive networks, the motions of all species tend to plateau. This phenomenon is considered to signify the formation of a gel. Figures are reproduced from Habasaki, J., Ishikawa, M. (2014), Molecular dynamics study of coagulation in silica-nanocolloid–water–NaCl systems based on the atomistic model, *Phys. Chem. Chem. Phys.*, 16, 24000–24017, with permission from the PCCP Owner Societies.

The MSDs of atoms and ions for system I-C are presented in Fig. 8.11b. In this figure, MSD was plotted before the equilibration of the system. The curve for each type of particle exhibits a plateau near the region of 100 ps after the start of the observation, and we conclude that gelation occurred at this time. After some time (~ns), the system returned to a more mobile state by bond breaking. The change can be reversible before a certain number of bonds being formed among the clusters.

As shown in Figs. 8.10 and 8.11, both changes in the solvent structure and the reduction of mobility occur accompanied with the coagulation. One may consider that the confinement in the networks or pores is a cause of the changes in the dynamics of water and others may consider that the slowing down of the water by the addition of salt causes the gelation. The arrest of motion in the system is found in parallel with the formation of gel networks and it is difficult to judge the causal dependence.

## 8.7 Other Approaches

Besides the fully atomistic simulations, there are several models to treat colloidal systems. In research of typical colloidal particles, which are generally larger than the solvent molecules, details of the structures of the solvents and colloids tend to be neglected. Under these circumstances, one typical approach to the study of colloidal systems is based on the mean-field approach, in which the solution is treated as a homogeneous medium. Only interactions among colloidal units are focused on such approaches.

However, in the treatment of nanocolloidal systems, the size of solvent molecules is not negligible. Therefore, careful treatment seems to be required.

Properties of the colloids depend on several factors and therefore further examination of the roles of the pH value and the surface of the colloid will be useful for understanding the behaviors of colloids [37, 38].

A more coarse-grained model of colloids might be used for some purposes if the roles of detail of colloidal structure were clear.

Simulations can also be done with some assumptions about how the coagulation process occurs. Several Monte Carlo (MC) simulations have been done using a sticking probability, $P_{ij}$. Using off-site Monte-Carlo simulations, Kim et al. [39] have accounted for the surface charges of the colloids in their examination of the effects of pH on the fractal structures of aggregates. In their work, the DLVO theory is used to provide the relation between the experimental process parameters and the $P_{\mathrm{ij}}$ value. Experimentally, it is known that silica surfaces in colloidal systems exhibit a rich variety of behaviors by the existence of siloxane or silanol, depending on the conditions of preparation such as pH. Surface modifications also affect the properties of colloids and may play important roles under some circumstances. Thus far, several types of parameters for expressing the potential between the silica surface and the water [40–42] have been investigated.

Hassanali and Singer [41] have proposed a set of parameters to describe the potential of a hydroxylated silica surface (based on the BKS model and SPC water) for $SiO_2$–water interactions. Reactive force fields, such as ReaxFF [42], have been used in several related works [42–46]. Such approaches will be useful in the investigation of the details of reactions on the surfaces of colloids or gels. Of course, the *ab initio* MD such as the Car–Parrinello (CP) method [47] will be a promising approach to considering the detailed properties of colloidal surfaces and coagulation processes, although this approach is not necessarily suitable for the examination of slow processes and large systems, because extensive calculation resources and long times are required.

## 8.8 Conclusion

In this chapter, we presented the results of full atomistic simulations of complex systems of colloidal silica and water, both for structures and dynamics of these systems with variations in the concentration of salt and/or the packing density of the colloids. Large system sizes were necessary to ensure that long length scales could be covered for the characterization of the gel, although actual system sizes might be limited due to the available computational resources and real time required for calculations. By using the potential model, which allows the breaking of the Si–O bond, the coagulation of colloids can be examined without assumptions of the mechanism of the formation.

By the fully atomistic models, a spontaneous formation of gel is observed. The process is found to be accompanied with the structural and changes in the dynamics of water and salts. It means that their dynamics and structures play a role in the coagulation process. A description of the structure of aqueous solution, including the positions of cations and anions, is necessary, although such factors tend to be neglected in some other approaches.

Systematic changes in $g(r)$ of Si–Si pairs with variations of concentrations of NaCl and $SiO_2$ contents are observed and the barrier height of the coagulation process is explained by the concept of potential of the mean force (PMF). The results obtained here will be beneficial for gaining a better understanding of aggregation and gelation in colloidal silica systems. The addition of NaCl causes changes in the structure and dynamics of the solvent, and the aggregation process is affected by these changes. NaCl not only causes changes near the surface of the colloid but also causes structural changes of the water. This phenomenon is closely related to a polyamorphism of water.

The results for colloidal solutions and gels presented here will be useful to understand how interactions among atoms work to form the nanoscale structures. In this chapter, a role and the importance of atomistic simulations in the study of complex systems is also clarified.

# Appendix

## 8.A.1 Potential Function and Parameters Used in Atomistic Simulations

The availability of reliable potential parameters and their consistency (including the consistency of functional forms) in complicated systems are problematic, although many efforts to derive such parameters are underway. In the present work, potential parameters were taken from the literature to maintain consistency of quality and combinations as much as possible. Here, we briefly explain the parameters used in the present work.

Tsuneyuki et al. [48] first developed potential parameters for silica (TTAM) based on *ab initio* MO calculations at the Hartree-Fock level. Several such sets of parameters can be obtained, due to the non-linearity of the functional form; the set that represents bulk properties was chosen because the distribution of charges in a small cluster is not the same as that in the bulk. This model is known to successfully reproduce many polymorphs of silica; therefore, it can represent many local structures with different Si–O distances and different O–Si–O and Si–O–Si angles.

In the present work, the TTAM model was used to model a silica colloid, in which the charges for Si and O were set to +2.4 and −1.2, as in the bulk model. The potential function of the TTAM model consists of a Coulombic term, a Gilbert–Ida-type [49] and an $r^{-6}$ term (for the correction of the softness of the oxygen atom.):

$$\phi_{ij} = \frac{z_i z_j e^2}{r} + f_0(b_i + b_j)\exp\frac{(a_i + a_j - r)}{b_i + b_j} - c_i c_j r^{-6} \tag{8.A.1}$$

Here the first term on the right-hand side represents the Coulombic interaction. This term was used for all combinations of atoms. The value of $r$ represents the distance between atoms, and $a_i$ and $b_i$ are the effective radius and the softness parameter, respectively, of atom i with a standard force $f_0$. The repulsive part of the function can be easily transformed into a Born–Meyer–Huggins (BMH)-type form.

$$\phi_{ij} = A\exp(B(\sigma - r_{ij})) \tag{8.A.2}$$

The parameters in (8.A.1) are additive, not only for pairs but also for atomic species. Therefore, the transferability of the parameters is good. Our potential parameters for alkali silicates [50] have been successfully used by many groups. The parameters used for the silica components of the parameters are common with the TTAM parameters. Therefore, by using this parameter set, a system with a glass plate can be treated with consistency. This feature is useful for the study of colloidal systems because recent experimental measurements have revealed that the contribution of long-range attraction is non-negligible for colloidal spheres confined by charged glass walls [51], although no such attraction has been observed in analogous measurements in unconfined systems.

This simple functional form using a small number of parameters can be used in many types of MD programs, with and without some modifications. During the aggregation and gelation processes investigated in the present work, the reconstruction of Si–O bonds was observed.

In the model, the charge-redistribution effect caused by the lengthening of bonds or changing angles was included in the parameters that represented the total potential surface, because in the MO calculations of the silica cluster used for modeling the Si–O distance and the bond angles could vary, and the charge distribution follows from these changes. Namely, the charges were fixed only in the fitted parameters, and hence, the TTAM potential could be used to represent bond breaking and bond formation. Since TTAM model was optimized for bulk material, applicability of it will be better in larger aggregates and gels than in a small cluster.

The repulsive parts of the SPC/E parameters are given by the Lennard–Jones (LJ) functional form:

$$\phi_{ij} = 4\varepsilon\left[\left(\frac{\sigma_{ij}}{r_{ij}}\right)^{12} - \left(\frac{\sigma_{ij}}{r_{ij}}\right)^{6}\right] \tag{8.A.3}$$

For the interaction between water and silica, the repulsive interactions were estimated using a combination rule. Before the rule was applied for the interaction with silica, the parameters associated with the LJ form were transformed [52] into the corresponding parameters for a Gilbert–Ida-type potential by requiring that the two functions have the same energy-minimum separation and depth and the same behavior in the large-$r$ region.

The interaction between silica and $Na^+$ or $Cl^-$ was obtained from the CLAYFF Force Field [12]. The following characters are not included in the present modeling of the system: details of surface structures such as the existence of Si–O–H and chemical effects such as dissociation of water. Some of graphics used in this chapter are prepared by using VESTA by Momma and Izumi [53].

## References


1. Habasaki, J., Ishikawa, M. (2014), Molecular dynamics study of coagulation in silica-nanocolloid–water–NaCl systems based on the atomistic model, *Phys. Chem. Chem. Phys.*, 16, 24000–24017.
2. Derjaguin, B., Landau, L. (1941), Theory of the stability of strongly charged lyophobic sols and of the adhesion of strongly charged particles in solutions of electrolytes, *Acta Phys. Chem. URSS*, 14, pp. 633–662.
3. Verwey, E. J. W., Overbeek, J. Th. G. (1948), *Theory of the Stability of Lyophobic Colloids*, Amsterdam: Elsevier.

4. Eslahian, K. A., Majee, A., Maskos, M. Würger, A. (2014), Specific salt effects on thermophoresis of charged colloids, *Soft Matter*, 10, pp. 1931–1936. 10.1039/C3SM52779D.
5. Wang, T.-Y., Sheng, Y.-J., Tsao, H.-K. (2006), Multivalent effects of multivalent salt addition on effective charge of dilute colloidal solutions, *J. Chem. Phys.*, 125, p. 194523 (1–11); https://doi.org/10.1063/1.2390707.
6. Lyklema, J. (2013), Coagulation by multivalent counterions and the Schulze–Hardy rule, *J. Colloid and Interface Sci*, 392, pp. 102–104.
7. Mishima, O., Calvert, L. D., Whalley, E. (1985), An apparently first-order transition between two amorphous phases of ice induced by pressure, *Nature*, 314, pp. 76–78.
8. Tsuneyuki, S., Tukada, M., Aoki, M., Matsui, Y. (1988), First Principles interatomic potential of silica applied to molecular dynamics, *Phys. Rev. Lett.*, 61, pp. 769–872.
9. Berendsen, H. J. C., Grigera, J. R., Straatsma, T. P. (1987), The missing term in effective pair potentials, *J. Phys. Chem.*, 91, pp. 6269–6271.
10. Mark, P., Nilsson, L. (2001), Structure and dynamics of the TIP3P, SPC, and SPC/E water models at 298 K, *J. Phys. Chem. A*, 105, pp. 9954–9960.
11. Joung I. S., Cheatham III, T. E. (2008), Determination of alkali and halide monovalent ion parameters for use in explicitly solvated biomolecular simulations, *J. Phys. Chem. B*, 112, pp. 9020–9041.
12. Cygan, R. T., Liang, J.-J. Kalinichev, A. G. (2004), Molecular models of hydroxide, oxyhydroxide, and clay phases and the development of a general force field, *J. Phys. Chem. B*, 108, pp. 1255–1266. DOI:10.1021/jp0363287.
13. Todorov, I., Smith, W. (2004), DL_POLY_3: The CCP5 national UK code for molecular-dynamics simulations, *Phil. Trans. R. Soc. Lond. A*, 362, pp. 1835–1852.
14. Nosé, S., Klein, M. L. (1983), Constant pressure molecular dynamics for molecular systems, *Mol. Phys.*, 50, pp. 1055–1076.
15. Nosé, S. (1984), A unified formulation of the constant temperature molecular dynamics methods, *J. Chem. Phys.*, 81, pp. 511–519.
16. Hoover, W. G. (1985), Canonical dynamics: Equilibrium phase space distributions, *Phys. Rev. A*, 31, 1695–1697.
17. Darden, T., York, D., Pedersen, L. (1993) Particle mesh Ewald: An N.log(N) method for Ewald sums in large systems, *J. Chem. Phys.*, 98, pp. 10089–10092.

18. Hasmy, A., Anglaret, E., Foret, M., Pelous, J., Jullien, R. (1994), Small-angle neutron-scattering investigation of long-range correlations in silica aerogels: Simulations and experiments, *Phys. Rev. B*, 50, pp. 6006–6016.

19. Kirkwood, J. G. (1935), Statistical mechanics of fluid mixtures, *J. Chem. Phys.*, 3, pp. 300–313.

20. Chandler, D. (1987), *Introduction to Modern Statistical Mechanics*, Oxford University Press, New York.

21. Hansen, J. P., MacDonald, I. R. (1990), *Theory of Simple Liquids,* 2nd ed., Academic Press, London.

22. Gray, C. G., Gubbins, K. E. (1984), *Theory of Molecular Fluids,* vol. 1, Clarendon Press, Oxford.

23. Shankar, K., Rosenberg, J. M., Bouzida, D., Swendsen, R. H., Kollman, P. A. (1992), The weighted histogram analysis method for free-energy calculations on biomolecules. I. The method, *J. Comput. Chem.,* 13(8) pp. 1011–1021. doi:10.1002/jcc.540130812.

24. Bartels, C. (2000), Analyzing biased Monte Carlo and molecular dynamics simulations, *Chem. Phys. Lett.,* 331(5–6), pp. 446–454. doi:10.1016/S0009-2614(00)01215-X.

25. Allen, T. W., Anderson, O. S., Roux, B. (2006), Molecular dynamics—potential of mean force calculations as a tool for understanding ion permeation and selectivity in narrow channels, *Biophys. Chem.,* 124, pp. 251–267. doi.10.1016/j.bpc.2006.04.015.

26. Keblinski, P., Eggebrecht, J., Wolf, D., Phillpot, S. R. (2000), Molecular dynamics study of screening in ionic fluids, *J. Chem. Phys.,* 113, pp. 282–291.

27. Habasaki, J., Ngai, K. L. (2011), Multifractal nature of heterogeneous dynamics and structures in glass forming ionic liquids, *J. Non-Cryst. Solids*, 357, pp. 446–453.

28. Poole, P. H., Sciortino, F., Essmann, U., Stanley, H. E. (1992), Phase behavior of metastable water, *Nature,* 360, pp. 324–328.

29. Liu, D., Zhang, Y., Chen, C. C., Mou, C. Y., Poole, P. H., Chen, S. H. (2007), Observation of the density minimum in deeply supercooled confined water, *Proc. Natl. Acad. Sci.,* 104, pp. 9570–9574.

30. Soper, A. K., Ricci, M. A. (2000), Structures of high-density and low-density water, *Phys. Rev. Lett.*, 84, pp. 2881–2884.

31. O. Mishima, Calvert, L. D., Whalley, E. (1985), An apparently first-order transition between two amorphous phases of ice induced by pressure, *Nature*, 314, 76–78.

32. Habasaki, J. (2019), Heterogeneous–homogeneous transition and anomaly of density in SPC/E water examined by molecular dynamics simulations, *Physica A*, 527, 121391.

33. Ngai, K. L., Habasaki, J., Hiwatari, Y., León, C. (2003), A combined molecular dynamics simulation, experimental and coupling model study of the ion dynamics in glassy ionic conductors, *J. Phys. C: Condensed Matter*, 15, S1607–S1632. doi:10.1088/0953-8984/15/16/310.

34. Ngai, K. L., Habasaki, J., Leon, C., Rivera, A. (2004), Comparison of dynamics of ions in ionically conducting materials and dynamics of glass-forming substances: remarkable similarities, *Z. Phys. Chem.*, 219, 47–70.

35. Habasaki, J., Ngai, K. L., Hiwatari, Y. (2002), Molecular dynamics study of cage decay, near constant loss, and crossover to cooperative ion hopping in lithium metasilicate, *Phys. Rev. E*, 66, 021205(1–11).

36. Habasaki, J., Ngai, K. L., Hiwatari, Y. (2004), Dynamics of caged ions in glassy ionic conductors, *J. Chem. Phys.*, 120, 8195–8200.

37. Tanaka, H., Araki, T. (2000), Simulation method of colloidal suspensions with hydrodynamic interactions: Fluid particle dynamics, *Phys. Rev. Lett.*, 85, pp. 1338–1341. 10.1103/PhysRevLett.85.1338.

38. Depasse, J., Watillon, A. (1970), Stability of amorphous. Colloidal silica, *J. Colloid Int. Sci.*, 33, pp. 430–438.

39. Kim, S., Lee, K.-S., Zachariah, M. R., Lee, D. (2010), Three-dimensional off-lattice Monte Carlo simulations on a direct relation between experimental process parameters and fractal dimension of colloidal aggregates, *J. Colloid Interface Sci.*, 344, pp. 353–361.

40. Cole, D. J., Payne, M. C., Csányi, G., Spearing, S. M., Ciacchi, L. C. (2007), Development of a classical force field for the oxidized Si surface: Application to hydrophilic wafer bonding, *J. Chem. Phys.*, 127, 204704(1–12).

41. Hassanali, A. A., Singer, S. J. (2007), Model for the water–amorphous silica interface: The undissociated surface, *J. Phys. Chem. B*, 111, pp. 11181–11193.

42. van Duin, A. C. T., Strachan, A., Stewman, S., Zhang, Q., Xu, X., Goddard, III, W. A. (2003), ReaxFF SiO reactive force field for silicon and silicon oxide systems, *J. Phys. Chem.*, 107, pp. 3803–3811.

43. Manzano, H., Moeini, S., Marinelli, F., van Duin, A. C. T., Ulm, F.-J. Pellenq, R. J.-M. (2012), Confined water dissociation in microporous

defective silicates: Mechanism, dipole distribution, and impact on substrate properties, *J. Am. Chem. Soc.*, 134, pp. 2208–2215.

44. Lockwood, G. K., Garofalini, S. H. (2009), Bridging oxygen as a site for proton adsorption on the vitreous silica surface, *J. Chem. Phys.*, 131, 074703(1–8).
45. Fogarty, J. C., Aktulga, H. M., Grama, A. Y., van Duin, A. C. T., Pandit, A. A. (2010), A reactive molecular dynamics simulation of the silica-water interface, *J. Chem. Phys.*, 132, 174704(1–10).
46. Tuinier, R., Rieger, J., de Kruif, C. G. (2003), Depletion-induced phase separation in colloid–polymer mixtures, *Adv. Colloid Interface Sci.*, 103, pp. 1–31.
47. Car, R., Parrinello, M. (1985), Unified approach for molecular dynamics and density-functional theory, *Phys. Rev. Lett.*, 55, pp. 2471–2474.
48. Tsuneyuki, S., Tsukada, M., Aoki, H., Matsui, Y. (1988), First-principles interatomic potential of silica applied to molecular dynamics, *Phys. Rev. Lett.*, 61, pp. 869–872. DOI: http://dx.doi.org/10.1103/PhysRevLett.61.869.
49. Ida, Y. (1976), Interionic repulsive force and compressibility of ions, *Phys. Earth Planet Interiors*, 13, pp. 97–104.
50. Habasaki, J., Okada, I. (1992), Molecular dynamics simulation of alkali silicates based on the quantum mechanical potential surfaces, *Mol. Simul.*, 9, pp. 319–326.
51. Croker, J. C., Grier, D. G. (1996), When like charges attract: The effects of geometrical confinement on long-range colloidal interactions, *Phys. Rev. Lett.*, 77, pp. 1897–1900.
52. Bakaev, V. A., Steele, W. A. (1999), On the computer simulation of a hydrophobic vitreous silica surface, *J. Chem. Phys.*, 111, pp. 9803–9812.
53. Momma, K., Izumi, F. (2011), VESTA 3 for three-dimensional visualization of crystal, volumetric and morphology data, *J. Appl. Crystallogr.*, 44, pp. 1272–1276.

## Chapter 9

# Nanostructures of Aggregates and Gels Formed by Fully Atomistic Molecular Dynamics Simulations

In the previous chapter, the spontaneous formation of the gel from complex systems consisting of the nanocolloidal-silica-water–NaCl was examined by molecular dynamics simulations using a fully atomistic model. In this chapter, details of the structure of the gel thus formed are examined and some physical properties such as pair correlation functions and fractal dimensions are discussed [1]. The calculated fractal dimension is comparable to the experimental ones for the aggregation of silica colloids explained in Section 3.2 of Chapter 3, if the differences in the dimensionality of the methods of observations are correctly considered. At longer length scales, super-aggregation was evident in the gelation process.

## 9.1 Multiple Network Structures Found in Gels

The gel discussed here was formed without any prior assumption of the underlying mechanism of the formation in classical

*Molecular Dynamics of Nanostructures and Nanoionics: Simulations in Complex Systems*
Junko Habasaki

ISBN 978-981-4800-77-8 (Hardcover), 978-1-003-04490-1 (eBook)
www.jennystanford.com

MD simulations as mentioned in ref. [1] and in Chapter 8 from nanocolloidal silica–water–salt system. The detail of MD simulations is also found in Chapter 8. As shown in Fig. 9.1, the silica gel thus spontaneously formed has multiple-networks (a network of silica (composed of $SiO_4$ units), that of water and that of ions) and therefore we focus on the structures of networks. These substructures with different characteristics have mixed each other to form complex structures of the gel. Here these structures are characterized by several methods such as fractal dimension analyses, dendrograms, and distribution of $Q_n$ (here $n$ means the number of bridging oxygen in $SiO_4$ units) structures.

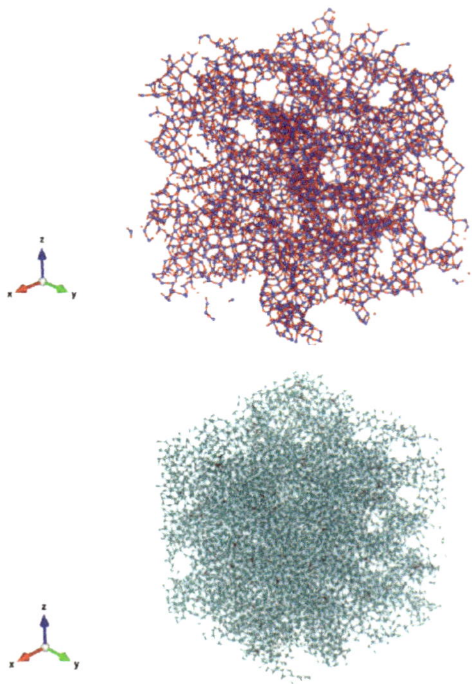

**Figure 9.1** (Top) Example of nanoporous structures of silica gel spontaneously obtained from nanocolloidal-silica—water–NaCl system. Structures are for a silica part. That is, water and NaCl are not shown in this figure. Red and blue particles are for oxygen and silicon, respectively. (Bottom) A water part of the same system as the upper one. The three-dimensional connections of porous structures are clear from this view. This structure corresponds to holes in gels. As shown in these figures, the structure of gel is formed by multiple networks including the salt part, which will be shown later.

As a first step to learn the characteristics of the coagulation process, a systematic examination of structures of systems with different magnitude of coagulations is done. In this section, several kinds of functions and analyses are used for characterizing structures of silica gels comparing with clusters and aggregates.

In Table 9.1, some characteristics of systems examined are listed with the images of networks (silica part) obtained by MD simulations. An image for I-C is after gelation. As summarized in Table 9.1, systems of series I have larger contents of silica compared with series II, where II-A is a reference system without salt. Of course, with the addition of NaCl, stronger coagulation is produced as shown in series I. Almost all of simulations were performed at 300 K, while higher temperature (400 K) was set in the system I-C. During the long run in that system, a spontaneous formation of the gel with volume reduction (~19%) was observed.

**Table 9.1** Characteristics of systems examined in Chapters 8 and 9

| System | Ratio of numbers of NaCl pairs and water molecules | Concentration of NaCl in water (mol%) | Temperature (K) | Status and network structures  |
|---|---|---|---|---|
| I-A | 128:8192 | 1.56 | 300 | Aggregate  |
| I-B | 320:8192 | 3.9 | 300 | Aggregate 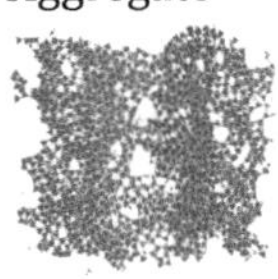 |
| I-C | 320:8192 | 3.9 | 400 | Gel  |

(*Continued*)

**Table 9.1** (*Continued*)

| System | Ratio of numbers of NaCl pairs and water molecules | Concentration of NaCl in water (mol%) | Temperature (K) | Status and network structures  |
|---|---|---|---|---|
| II-A | 0:15872 | 0 | 300 | Clusters 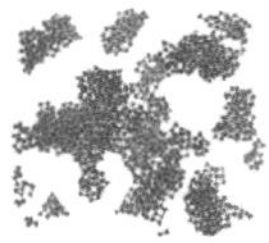 |
| II-B | 160:15872 | 1.01 | 300 | Clusters and Aggregate 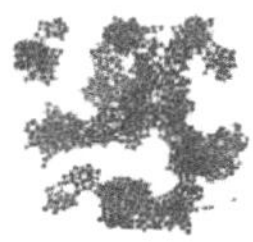 |
| II-C | 320:15872 | 2.02 | 300 | Aggregate 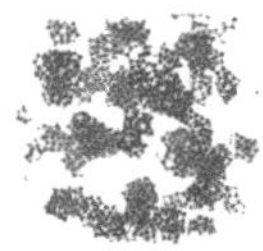 |

## 9.1.1 Structure of the Silica Part

For the examination of networks formed by $SiO_4$ units, pair correlation functions, fractal dimension analyses, $Q_n$ structures and dendrogram are used here. Fundamentals of these methods are discussed in Chapter 4. Before showing new data, some of the results discussed in Chapter 8 are briefly summarized here.

### 9.1.1.1 Pair correlation functions of Si–Si pairs: characteristics of silica part in clusters, aggregates, and gels

First, the structures of the networks of the silica part will be characterized by pair correlation functions, $g(r)$s, for the Si–Si pairs.

In Fig. 9.2, systematic changes of the pair correlation functions for the Si–Si pairs of systems I-A, I-C, II-A, and II-C are shown using a log-log scale. The series of water-rich systems (series II) exhibit higher peaks than the series of silica-rich systems (series I). The former indicates more isolated clusters.

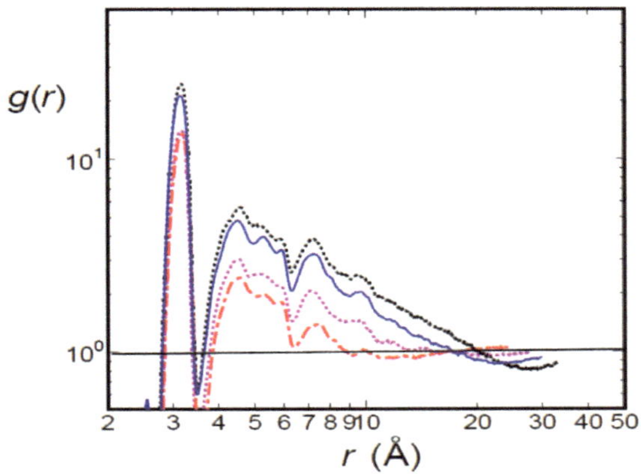


**Figure 9.2** Comparison of pair correlation functions, $g(r)$ of Si–Si pairs. From upper to lower in the second peaks are for II-A, II-C, I-A and I-C (gel), respectively. The upper two curves (water-rich systems) have clearer minima at a longer length scale meaning the separation among clusters or aggregates. In the system with the formation of gel, the network is spreading, and the minimum becoming small and shifting to the lower $r$ region. Reproduced from Habasaki, J., Ishikawa, M. (2014), Molecular dynamics study of coagulation in silica-nanocolloid–water–NaCl systems based on the atomistic model, *Phys. Chem. Chem. Phys.*, **16**, pp. 24000–24017. With permission from the PCCP Owner Societies.

The following characteristics with the development of coagulations are found:

1. The small minimum (negative deviation from $g(r) = 1$) is observed in the long-length-scale regions ($r > 12{\sim}35$ Å) of the dilute systems before $g(r) = 1$ is attained. The largest deviation from the line for $g(r) = 1$ is observed in the system without salt (II-A) at approximately $r = 30$ Å. The positions of the minima are located at approximately half of the mean distance between clusters or aggregates and it means that these structures are separated by solvent (and salt). The size of these clusters is several times larger than that of the original $SiO_2$ colloidal units in our systems.
2. With the development of the coagulation, both the position and depth of the minimum tend to decrease and finally, gels are formed as found in system I-C.

A comparable existence of the minimum is found by off-lattice diffusion limited cluster aggregation (DCLA) calculations performed by Hasmy et al. [2].

The change of the minimum can characterize the magnitude of the coagulation processes because it corresponds to the maximum of the potential of the mean force (PMF) [3] as discussed in Chapter 8. That is, a deeper minimum corresponds to a higher barrier height for the coagulation process.

#### 9.1.1.2 Coexistence of two length scale regions

Another important finding in $g(r)$ of Si–Si pair is the coexistence of two length scale regions, especially in the gel.

A decrease in the peak heights of $g(r)$ in the short-length-scale region can be observed with increasing NaCl concentration; hence, changes occur with coagulation. The system I-C can be well characterized by the existence of super-aggregates. In I-C after gelation, the power-law region below $r \sim 12$ Å represents the spreading of the network of clusters, and the longer-length-scale region represents the development of connections among these structures to form a gel.

To characterize the networks of the silica part, $Q_n$ distribution ($n$ is the number of bridging oxygens for $SiO_4$ units) has been widely used in glasses and crystals both in experiments [4–8] and in MD simulations [9] including silicates. The $Q_n$ (notation $Q^n$ is also used) structure means the $SiO_4$ unit with $n$ bridging oxygens.

Some figures for distributions are shown in Chapter 8, while in Table 9.2, the results of $Q_n$ distributions are shown with mean numbers of bridging oxygens and equilibration constant, $K$.

Both distribution and the mean number of bridging oxygens do not show remarkable changes by systems having a different magnitude of coagulation.

These results mean that the coagulation process proceeds by the rearrangement of Si–O bonds with keeping a total number of bonds with small changes in $Q_n$ distributions as a first approximation. In aggregates and gels, Some $Q_5$ structures (for $SiO_5$ units) are observed, while the contribution of $Q_2$ structure tends to decrease. Non-tetrahedral coordinated Si has been observed experimentally in several systems. For example, Stebbins and McMillan [10] observed five and six-coordinated Si by using $^{29}$SiMASNMR.

**Table 9.2** $Q_n$ distribution (in %) for systems shown in Table 9.1 with the mean number of bridging oxygens for the network of silica part

| System | Solvated-unit | System I-A | System I-B | System I-C | System II-A | System II-B | System II-C |
|---|---|---|---|---|---|---|---|
| Status | | Aggregate | Aggregate | Gel | Clusters | Aggregate | Aggregate |
| $Q_1$ | — | — | 0.24 | — | — | | — |
| Q2 | 3.18 | — | 3.17 | 1.14 | 0.32 | 0.37 | — |
| Q3 | 18.94 | 15.58 | 25.98 | 21.93 | 15.63 | 13.74 | 13.81 |
| Q4 | 75.76 | 78.64 | 68.84 | 75.19 | 79.76 | 80.44 | 80.55 |
| Q5 | 0 | 5.62 | 1.77 | 1.74 | 4.29 | 5.45 | 5.58 |
| Mean number of bridging oxygens | 3.7 | 3.9 | 3.7 | 3.8 | 3.9 | 3.9 | 3.9 |
| $K = \langle Q_4 \rangle^2 / (\langle Q_3 \rangle \langle Q_5 \rangle)$ in % | — | 70.63 | 103.06 | 148.16 | 94.88 | 86.41 | 84.20 |

*Source*: Reproduced with a small modification from Habasaki, J., Ishikawa, M. (2014), Molecular dynamics study of coagulation in silica-nanocolloid–water–NaCl systems based on the atomistic model, *Phys. Chem. Chem. Phys.*, 16, pp. 24000–24017, with permission from the PCCP Owner Societies.

Only small changes were observed in series I as shown in Chapter 8. However, the difference becomes clearer if we use the equilibrium (or quotient) constant, $K$ of the reaction among $Q_n$ structures.

In the bottom of Table 9.2, equilibrium (or quotient) constant, $K$ of the reaction,

$$Q_3 + Q_5 \Leftrightarrow 2Q_4$$

is shown, with tentatively neglecting $Q_1$ and $Q_2$ structures. For the gel, $K$ the value is found to be larger than other systems. Some difference is observed in $K$. The larger value of $K$ in the gel means the development of homogeneous networks concerning the structural units. As found in these values, $K$ can be a sensitive index to represent the small changes in the distribution and resultant networks. This suggests that the gel is rather homogeneous than other systems with aggregates or clusters, although further investigation may be required to establish the differences among distributions.

It is interesting to note that the distribution of $Q_n$ structures is closely related to the density of the system through the partial volume of the units [8, 9]. This is because the volume of the bridging oxygen is shared between two units. Namely, partial volume with larger $n$ value tends to be smaller. Due to these characteristics of different partial volumes, $Q_n$ distributions are also closely related to the polyamorphism [11] of the system in silicate glasses [9]. Because of the similarity of the shape of the structural units for silica (or silicates) and water including its phase-diagrams [11–13], it is interesting to compare polyamorphism in silica, silicates, and related systems with that of water based on the similar concepts of distribution of $Q_n$ structures [14] and its relationship with the heterogeneity of networks.

### 9.1.2 Fractal Dimensions of the Network Structures

Fractal dimension analyses are often applied to experimental observations; however, some caution is necessary for comparison among works, especially for systems with different observation methods.

For example, fractal dimension observed in experimental methods such as that from the TEM image might be different from the value obtained from pair correlation functions. The former is concerned with the two-dimensional images, while the latter is concerned with the radial direction data but can reflect the characteristics of three-dimensional structures. The value also depends on the length scale region where the dimension is determined. To clarify these problems, both two dimensional and three-dimensional methods are applied for the same systems obtained from molecular dynamics simulations in this section.

#### 9.1.2.1 Comparison of two- and three-dimensional fractal dimensions

Fractal dimensions were determined using 2-D projections of structures. In Fig. 9.3, an example of the two-dimensional projection

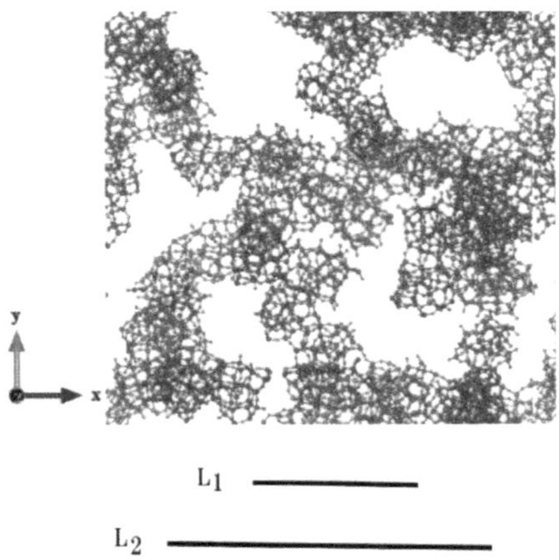


**Figure 9.3** Structure of clusters found in system II-A (the system without NaCl) for the basic cell of MD projected onto the *xy* plane. (Si and O atoms are represented by spheres). Length scales used in fractal dimension analysis of 2-D projection are shown on the bottom of the figure. The values for the medium-length-scale region were obtained using boxes with side lengths shorter than $L_1$, whereas those for the longer-scale region were obtained using boxes with side lengths between those of $L_1$ and $L_2$ in the figure. Although the structure looks like an aggregate from 2-D projection, nineteen clusters are evident in the 3-D view. Reprinted from Habasaki, J., Ishikawa, M. (2014), Molecular dynamics study of coagulation in silica-nanocolloid–water–NaCl systems based on the atomistic model, *Phys. Chem. Chem. Phys.*, 16, pp. 24000–24017., with permission from the PCCP Owner Societies.

of the network structures of silica in the MD cell is shown for the system II-A. In this system, 19 clusters are found from the three-dimensional view. These clusters are separated by the solvent but appear like a connected fractal structure in the 2-D projection. Fractal dimensions for the short- and long-range regions of the 2-D were obtained by the box-counting methods [15]. The length scales used for distinguishing two regions are shown in the figure. Fractal dimension thus obtained from 2-D projections are listed in Table 9.3.

**Table 9.3** Fractal dimensions of the colloidal systems with different magnitudes of coagulation, by 2-D and 3-D methods

| System | System I-A | System I-B | System I-C | System II-A | System II-B | System II-C |
|---|---|---|---|---|---|---|
| Status | Aggregate | Aggregate | Gel | Small clusters | Small clusters and aggregate | Aggregate |
| 2-D value Short length | 1.83 | 1.79 | 1.80 | 1.80 | 1.78 | 1.81 |
| 2-D value Long length | 2.00 | 1.17 | 1.43 | 1.17 | 2.00 | 1.17 |
| 3-D value Short length | 2.40 | 2.23 | 2.25 | 2.14 | 1.99 | 2.12 |
| 3-D value Long length | 2.58 | 3.07 | 3.01 | 2.10 | 2.34 | 2.40 |

*Source*: Reproduced from Habasaki, J., Ishikawa, M. (2014), Molecular dynamics study of coagulation in silica-nanocolloid–water–NaCl systems based on the atomistic model, *Phys. Chem. Chem. Phys.*, 16, pp. 24000–24017, with permission from the PCCP Owner Societies.

Another example of the structure, an aggregate found in system II-C, is shown in Fig. 9.4 in both 3-D and 2-D images. The dimension for the shorter length ($L < L1$) is due to local structure, while the longer length scale ($L1 \leq L < L2$) is related to the connection among these local structures. The latter value is smaller than the former when the clusters are separated by the solvent.

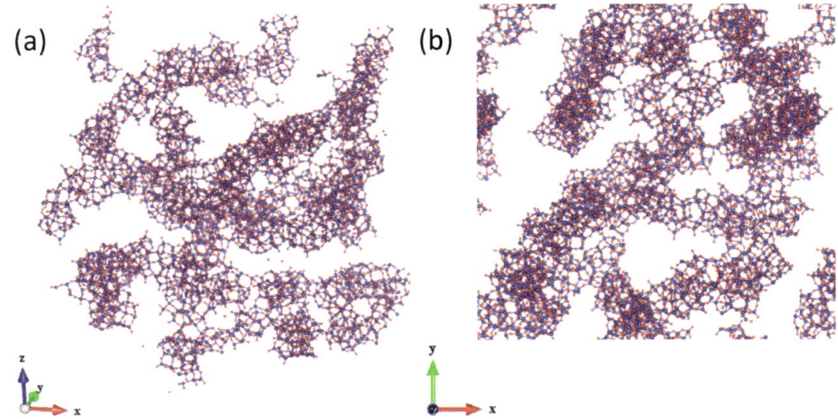


**Figure 9.4** Structure of clusters found in system II-C in 3D view and (b) that projected onto the *xy* plane. In this system, aggregate developed. Reprinted from Habasaki, J., Ishikawa, M. (2014), Molecular dynamics study of coagulation in silica-nanocolloid–water–NaCl systems based on the atomistic model, *Phys. Chem. Chem. Phys.*, 16, pp. 24000–24017, with permission from the PCCP Owner Societies.

Interestingly, for most of our systems, values near $d_f \sim 1.8$ were obtained by 2-D projections. The results mean that the 2-D values are not sensitive to the differences among the systems.

This is because the clusters tend to overlap in the images as shown in Figs. 9.3 and 9.4. If the clusters are connected by a limited number of (~1) bond, the dimension of ~1 is observed for a longer length scale value, whereas when the clusters are connected by several Si–O–Si bonds, the fractal dimension is ~2, which is the limiting value for the 2-D projection. It is interesting to compare this result for related experimental observation.

#### 9.1.2.2 Fractal dimension obtained from 3-D structure

The fractal dimension [15–19] of the structures is useful to characterize the networks in clusters, aggregates, and gels. One can determine the 3-D value of the fractal dimension for the same three-dimensional structures used for 2-D analyses by using MD simulations [1]. The three-dimensional fractal dimension of the network, $d_f$, can be defined by

$$d_f = \frac{d \log N(r)}{d \log r} \tag{9.1}$$

using the running coordination number $N(r)$.

From the relation for the structure factor [18],

$$S(q) \sim q^{-d_f}, \tag{9.2}$$

the fractal dimension can be obtained by

$$d_f = -\frac{d \log S(q)}{d \log dq} \tag{9.3}$$

The running coordination number $N(r)$ of species $j$ around species $i$ was also obtained from the MD simulations. The function $N(r)$ is related to the pair correlation function $g(r)$ as follows [20, 21] when $i$ and $j$ are the same species:

$$N(r) = <4\pi r^2 \sum_n \frac{N_j - 1}{V} g_{ij}(n\Delta r)>, \tag{9.4}$$

where $n$ is chosen such that $n\Delta r$ becomes $r$, and < > indicates an ensemble average.

A power law region is observed in the log-log plot of $N(r)$ and a shorter $r$ region for 3-D methods corresponds to the power law region of pair correlation function $g(r)$.

Although the differences among the systems can be observed in $g(r)$, the value obtained from $N(r)$ appears to be more accurate because it is based on the summation in Eq. (9.4) with better statistics.

Three-dimensional fractal dimensions of the Si aggregates (with and without bonding) were thus obtained from the slopes of the log-log plots in the power-law region [1], also listed in Table 9.2.

The fractal dimension, $d_f$, of the long-range structure of the aggregates can be determined from the slope of $N(r)$ in the region corresponding to $g(r) \sim 1$ after the power law region. In principle, one can use the region up to $\sim L/2$ for the estimation of the fractal dimension in the MD simulation. As shown in Table 9.2, from the 3-D values, one can clearly distinguish the clusters, aggregates, and gel.

In systems I-A, I-B, and I-C, we observe fractal dimensions of 2.2~2.4 in the shorter-length-scale region. The fractal dimension $d_f$ of I-A is slightly larger than that of I-B or I-C, whereas those for II-A and II-C are comparable. These values reflect a spreading out of the original colloidal units in the solvent but do not directly indicate the development of new connections among units.

A value of 2.53 is known to be the percolation threshold for a 3-D system [22] in the case of random occupation for a three-dimensional cubic lattice; the observed values seem to be slightly smaller than this typical threshold value. On the other hand, in the longer-length-scale region after the minimum ($r > 12{\sim}35$ Å), an increase in the fractal dimension is observed for higher NaCl concentrations. The fractal dimension obtained in this region is 2.58 for I-A and ~3.0 for I-B and I-C. Namely, systems I-B (just before gelation) and I-C exhibit similar exponents that indicate three-dimensional connections. The $g(r) \sim 1$ region is used to investigate the connection of colloidal units or clusters to form a gel, as mentioned above. In the case of I-C, the aggregates are thus three-dimensionally connected in the long-length-scale region, whereas the dimension in the shorter-length-scale region is slightly smaller than the percolation threshold, probably because of the existence of solvent-filled voids. Because different fractal dimensions (exponents) are observed at the two length scales, the system can be multifractal [15, 18, 19]—in other words, possessing several exponents—rather than monofractal.

#### 9.1.2.3 Comparison of fractal dimensions obtained by MD and those obtained in experiments

Using transmission electron microscopy (TEM) images, Weitz et al. have identified two distinct regions of aggregates with different fractal dimensions [23, 24]. They observed fractal dimensions of approximately $d_f \sim 1.8$ for diffusion-limited colloid aggregation and $d_f \sim 2.1$ for reaction-limited colloid aggregation, and the results were rather universal.

In the present work, for most of our systems, values near $d_f \sim 1.8$ were obtained (see Table 9.2) from 2-D images, whereas different values representing different magnitudes of coagulation of the systems were found in fractal dimension analysis of 3-D data.

The results indicate that the 2-D values are not sensitive to the differences among the systems. This situation can be understood by images of 2-D projections, where one cannot distinguish clusters and aggregates easily.

Here we consider more closely for the cause of rather general results in the 2-D methods.

One may expect that the projection onto a 2-D plane should have an exponent that is comparable to the corresponding 3-D value if the fractal dimension is less than 2. However, this was found not to be the case when the fractal dimension was greater than 2 in the 3-D observation. Projection onto a two-dimensional plane leads to a considerable degree of overlap of the clusters in the silica network and the voids (see Fig. 9.4). As a result, fractal dimension obtained becomes always less than the 2, which is the limiting value of the 2-D method.

Experimentally, Schaefer et al. [25] have determined the fractal geometry of colloidal aggregates of silica. From the structure factors (see Eqs. 9.2 and 9.3) obtained from both light and X-ray scattering, they obtained a value of $d_f = 2.12 \pm 0.05$, where a fractal structure of silica was observed in the region of 500~5000 Å. Our values for systems II-B and II-C are comparable to this value. In our systems, the region corresponding to $d_f \sim 2$ extends to shorter lengths because of the smaller size of the colloidal units. As a conclusion for the comparison of 2-D and 3-D methods, one can say fractal dimensions of aggregates or gels may not be obtainable from the analysis of 2-D images and further discussion should be based on the 3-D values.

From our study by using MD simulations, caution is necessary for the following facts in simulations of nanostructured materials. In complex systems with many components, longer length scales with different fractal dimensions might be found. If the system size was too small, the longer-length-scale region cannot be found in MD.

## 9.2 Multifractal Structures in Silica Gels: Coexistence of Different Length Scales

How do these fractal dimensions obtained by the 3-D method correlate with the gelation process? A value of 2.53 is known to be the percolation threshold for a 3-D system [22] in the case of random occupation for a three-dimensional cubic lattice; the observed shorter length scale values seem to be slightly smaller than this typical threshold value. On the other hand, in the longer-length-scale region after the minimum ($r$ > 12~35 Å), an increase in the fractal dimension is observed for higher NaCl concentrations. The fractal dimension is 2.58 for I-A and ~3.0 for I-B and I-C. Namely, systems I-B (just before gelation) and I-C exhibit similar exponents that indicate three-dimensional connections. The $g(r) \sim 1$ region is used to investigate the connection of colloidal units or clusters to form gel, as mentioned above. In the case of I-C, the aggregates are thus three-dimensionally connected in the long-length-scale region, whereas the dimension in the shorter-length-scale region is slightly smaller than the percolation threshold, probably because of the existence of solvent-filled voids.

Because different fractal dimensions (exponents) are observed at the two length scales [18, 19], the system can be multifractal [26–28]— in other words, possessing several exponents and are mixing — rather than monofractal nor simple mixtures.

As shown in Chapter 4, multifractality is often observed in other systems and seems to be a key feature to characterize complex systems. The multifractality of the structure of gels is called as super-aggregates argued in the following section.

### 9.2.1 Formation of Super-Aggregates and Multifractal Structures

The existence of different length scale regions is consistent with the findings in other theoretical calculations using some assumptions for the mechanism of the formation. Fry et al. [19] have reported a structural crossover between dilute regions of clusters and cluster–cluster aggregation (**super-aggregation**) regions based on off-lattice Monte Carlo simulations near gelation, assuming both reaction-limited cluster aggregation and ballistic cluster aggregation. In the dense region, the system possessed a fractal dimension of $d_f \sim 2.6$ over large length scales, whereas at smaller length scales, the system exhibited the limiting value associated with a dilute state.

Such structures are also found in experiments. Sorensen et al. [29] have reported structure factors $S(q)$ with large slopes in the small-$q$ regime in the aerogelation of a flame soot aerosol, indicating the existence of the further aggregation of clusters in a dense region as shown in Chapter 3. Our results, which were obtained without making any *a priori* the assumptions regarding the types of aggregation processes that would occur in the system, are consistent with these previous results.

One of the possible methods to examine the multifractal structures follows the length scales based on the charge density correlation function explained in Chapter 4.

The structures of the ionic systems are known to consist of the following two contributions, $Q(r)$ and $G(r)$, as known in other ionic systems [see Chapter 4]:

$$Q(r) = Q_{+}(r) + Q_{-}(r) = g_{++}(r) + g_{--}(r) - 2g_{+-}(r) \tag{9.4}$$

$$G(r) = (G_{+}(r) + G_{-}(r))/2 = (g_{++}(r) + g_{--}(r) + 2g_{+-}(r))/4 \tag{9.5}$$

The former is related to the structures of layers of charges (charge density wave, CDW), with a suitable weight of charges while the latter is related to structures of ionic positions regardless of the charges (density wave, DW). The structures of Si–Si pairs are formed by alternative charge-charge and density-density correlations of Si atoms and O atoms. Therefore, both the repulsive force and long-ranged Coulombic force contribute to the structural formation of clusters or aggregates. Naturally, the power law region of $g(r)$ was also found in other atom–atom pairs such as O–O, with comparable exponents.

Complicated structures with several length scales are also found in the structure of the salt in the system.

However, some cautions are required for the application of this method for complicated nanostructured materials. The method is useful for determining the length scale in the short-range even in the case of colloidal solutions and the system showing gelation at least qualitatively. For example, for the silica part, one can obtain $Q(r)$ as in the monovalent systems, if the definition of $Q(r)$ is slightly modified by including the number ratio of Si and O. For a longer scale, the following problems should be considered:

1. In decay-type functions such as $\ln|Q(r)r|$, it is not easy to determine meaningful slopes for long length scales.

2. Especially for the minor component, it is not easy to obtain statistically meaningful values.
3. Oscillation of the functions becomes a complicated one and it is not easy to distinguish regions to be examined.

Of course, for this purpose, one should use the large MD cell and long runs so that better statistics can be obtained. Otherwise, longer length correlations are simply missing or modified, and this may affect the values for short-length-scale motions and structures. A similar situation is found even in the simple water system if a small size of the simulation box is used.

### 9.2.2 Classifications of Clusters, Aggregates, and Gels by Multifractality

As shown in Chapter 2, the classification based on the fractal nature as well as the multifractal nature of the system is possible. If the fractal dimension of the longer length structure of the silica network determined from the running coordination number of Si–Si pairs is plotted against the fractal dimension of the shorter length, the structures obtained for I-A~I-C and II-A~II-C spread on the plot in Fig. 9.5 and the position on the plot characterizes the system well. The characteristics of systems are also shown in the figure.

When the $d_{\text{f-long}}$ value increases and exceeds ~2.5, which is comparable to the threshold value for the 3-D [22], gel seems to be formed. The value of $d_{\text{f-long}}$ changes in the order cluster<aggregate <gel, and the connectivity of long-ranged structures is well represented by $d_{\text{f-long}}$, while the values for $d_{\text{f-short}}$ tend to show fluctuation. However, if we exclude the system II-C with the large fluctuation of density (lacunarity), the value tends to become smaller with the increase of the content of the salt in each series. It seems to be a natural trend because the total number of colloidal units is the same in all systems.

This kind of plot can distinguish the heterogeneity (multifractality) of systems by the coexistence of short- and length-scale structures. The concept is not the same for the heterogeneity of the network, where different kinds of $Q_n$ structures can contribute.

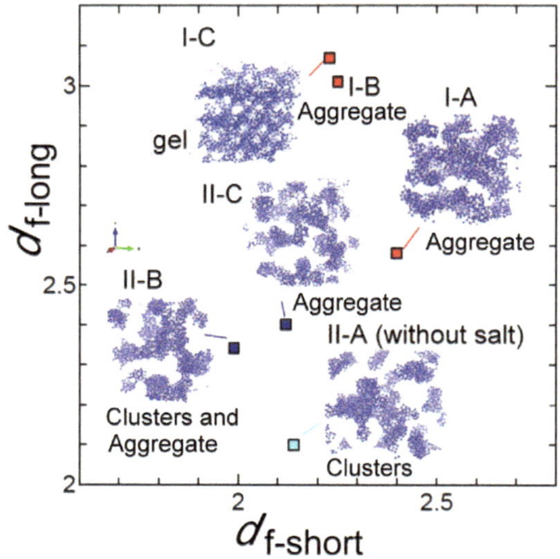


**Figure 9.5** Structures of silica part for clusters, aggregates and gel obtained from Molecular Dynamics Simulations of silica–nanocolloid–water–NaCl systems. Fractal dimensions of a longer length scale are plotted against fractal dimensions of short length scale regions for these structures. Then these systems can be characterized by the combinations of these fractal dimensions. Series II (blue squares) are more water-rich than series I (red squares). II-A (a pale blue square) is a reference system without salt. Contents of the salt are in the order of A<B (before gelation) = C (after gelation) in series I, while it is in the order of A<B<C in series II. Coagulation developed in the order of II-A<II-B<II-C<I-A<I-B<I-C. The development of the coagulation process can be represented by the position on this plot. Therefore, this kind of plot can be applied for expanded classification of complex systems as already discussed in Chapter 2.

## 9.3 Dendrogram of the Network Connection

To understand the mechanisms of the formation of aggregates and super-aggregates, further analysis of the bond formation and the tree structure (dendrogram) was performed as shown in this section.

Representation of the structures of coagulated systems by the dendrogram is one of the possible ways to examine the connections in networks.

The fundamentals of the methods are shown in Chapter 4 with a simple example. As explained there, the vertical axis of the dendrogram represents the distance or dissimilarity between clusters. Each connection of two clusters is represented by two veritical lines joining into one.

For the basis of the cluster analysis to obtain the dendrogram for colloidal systems, $SiO_4$ units are used. Examples of clusters of units are shown in Fig. 9.6 for the initial structure of series I systems and that for the gel in the I-C system. Each $SiO_4$ units is shown as a sphere.

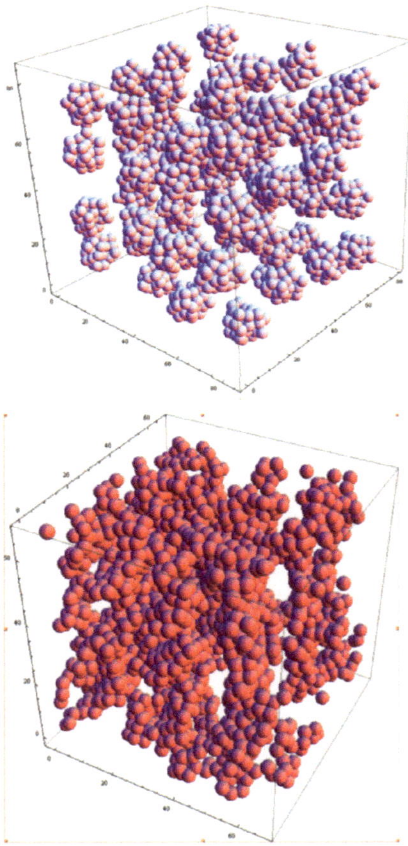

**Figure 9.6** (Top) Initial positions of $SiO_4$ units (spheres) in series I. (Bottom) Positions of $SiO_4$ units in a gel. Data for the positions of central Si are used to obtain the dendrogram shown in Fig. 9.9.

In Figs. 9.7, 9.8, and 9.9, dendrograms of clusters in II-A (clusters), II-C (aggregates), and I-C (gel) are shown, respectively.

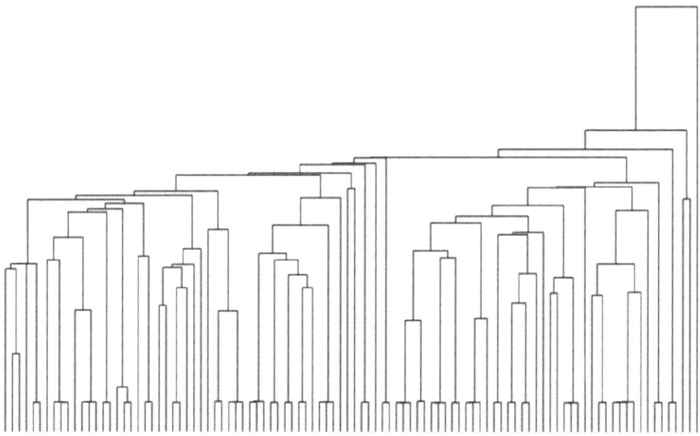

**Figure 9.7** An example of the dendrogram for the cluster structure of the system II-A. The diagrams are shown for the top 100 combinations. The nearest-neighbor distance (Euclidean distance) between Si atoms was used to quantify the degree of clustering. Small clusters are separated from other clusters as shown by the large distance between clusters.

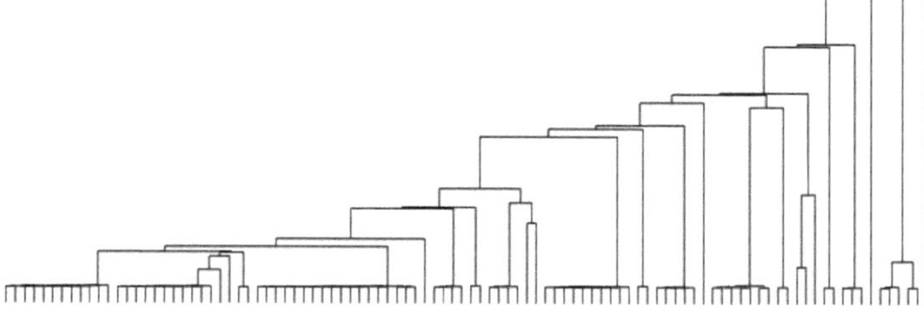

**Figure 9.8** An example of the dendrogram for the cluster structures of system II-C. The diagrams are shown for the top 100 combinations. The nearest-neighbor distance (Euclidean distance) between Si atoms was used to quantify the degree of clustering. In this case, larger aggregates seem to be formed by the combination of small clusters. However, the long distance among clusters will prevent further connections.

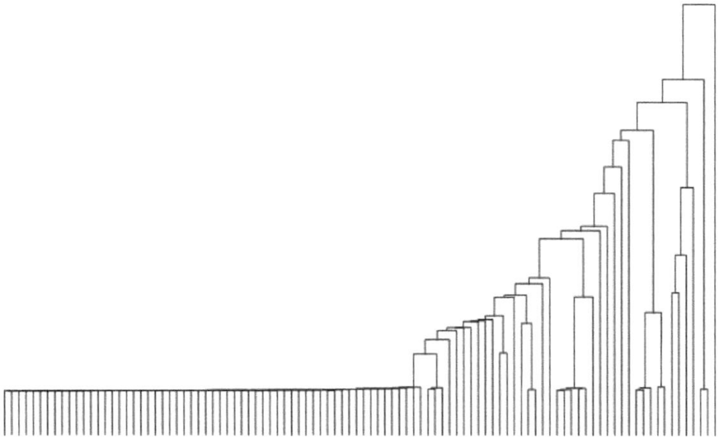

**Figure 9.9** An example of the dendrogram for the network structures of the system I-C (gel). The diagrams are shown for the top 100 combinations. The nearest-neighbor distance (Euclidean distance) between Si atoms was used to quantify the degree of clustering. In this case, smaller units are incorporated into the larger aggregate to form a gel. The pattern suggests that connection with nearby units occurred to form a large chain or aggregate and then other units were added to it.

In these figures, clustering was judged by the Si-(O)-Si distance. The top 100 combinations were selected for visibility. The nearest-neighbor distance (Euclidean distance) between Si atoms is used to quantify the degree of clustering.

In principle, the bottom row of nodes can represent individual units (Si in $SiO_4$ units in this case), and the remaining nodes represent the clusters to which the unit (Si) belongs, with the height of each connecting line representing the distance (dissimilarity).

In Fig. 9.7, the structure of system II-A is shown. As shown by the long distance in the vertical axis, clusters are separated by solvent and structures of clusters seem to be rather like each other.

In Fig. 9.8, the dendrogram for the network structures of system II-C is shown. In this case, small clusters seem to gradually combine into larger aggregate, and further aggregation seems to be suppressed by the larger separation among clusters at least in the condition of the static placement of the system. In experimental conditions, convection will help the development of aggregation.

In Fig. 9.9, the tree structure of the network in system I-C (gel) is presented. In the left-hand part of the dendrogram, a large aggregate with chain-like connections is evident. In other regions, smaller aggregates are present. Connections among aggregates of different sizes are also apparent. These findings are consistent with the formation of "super-aggregates" in the gelation process.

Heterogeneous development (with differently sized clusters) during coagulation is found, although one needs to be careful about the differences in length scales, where the heterogeneity is argued.

Because the clustering was quantified in terms of the nearest-neighbor distance in this analysis, these tree structures seem to have formed through the random connection of neighboring structures. This type of analysis will be useful in distinguishing the mechanisms of the formation of aggregates and gels.

In both Figs. 9.7 and 9.8, the structures seem to be rather heterogeneous during the structure formation. In the case of Fig. 9.7, the large aggregate is separated from other units and therefore, the structure formation should be slow, while in Fig. 9.8, the aggregate tends to spread over so that the distance from other units is not so large. Therefore, from MD simulations, both the roles of distances among developed clusters and barriers can be examined.

Cluster–cluster distances determined by different kinds of centers also can be used to describe further aggregation. Careful arguments for such results are required since a different impression might be obtained when a different kind of judgment for clustering is used, and hence one needs to choose the suitable criterion of the clustering according to the purpose of analyses.

## 9.4 Structure of the Water Part in Gels

### 9.4.1 Historical Views for High Density and Low Density Water

Before discussing the structure of water in colloidal solutions and gels, it may be better to describe the history of research of the polyamorphism of water briefly.

In 1984, Mishima and coworkers [30] reported the formation of high-density amorphous ice (HDA) under the pressure at 77 K. This ice was different from the low-density amorphous ice (LDA), which was obtained by the rapid cooling of water or by the deposition of the vapor. The change between LDA and HDA seems to be the first-order transition with the changes of ~20% in volume. The existence of different kinds of amorphous ice is called "polyamorphism."

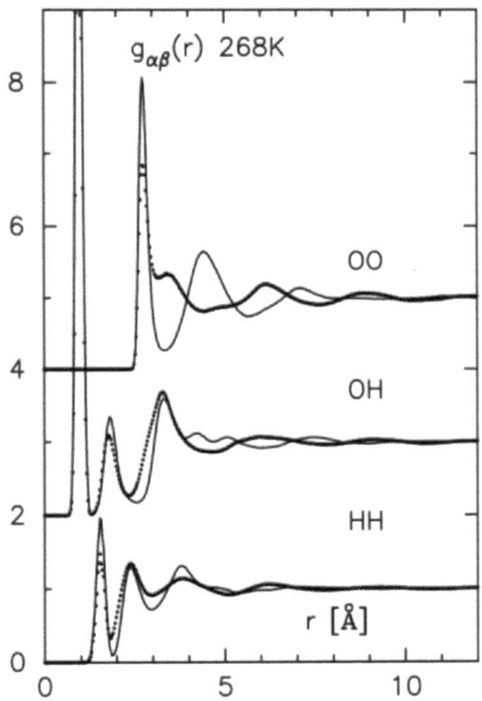


**Figure 9.10** Pair correlation functions of high-density water and low-density water. Reprinted from Soper, A. K., Ricci, M. A. (2000), Structures of high-density and low-density water, *Phys. Rev. Lett.*, 84, pp. 2881–2884, with permission.

In 1992, Poole and coworkers [31] performed the MD simulation of supercooled water at around 228 K. Anomalies of water such as density maximum at 4 °C and the minimum in the isothermal compressibility at 46 °C are thought to reflect the presence of singularities in the behavior of thermodynamic quantities occurring in the supercooled region. They assume the existence of a critical point for the liquid–liquid transition based on the "stability–limit conjecture." Below this critical point, two kinds of liquids can exist, that is, phase separation of high-density liquid (HD) and low-density liquid (LDL) occurs.

Many experiments and simulations have been tried to prove this hypothesis, although arguments are continuing.

Experimentally, Soper and Ricci [32] examined the pressure-dependent water structure at 268 K by the neutron diffraction. They found a transformation from low-density water (solid curves) and high-density water (dotted curves) as shown in Fig. 9.10. The former has been characterized by an open, hydrogen-bonded tetrahedral structure, while the latter in nontetrahedral O–O–O angles and a collapsed second coordination shell. That is the shell at around 5 Å was changed to the inner shell at around 4.2 Å. It is worthy to note that changes in further shells and changes in the sparse region among shells are also necessary to consider for a detailed argument for the density of the system [14].

### 9.4.2 Changes of Water Structures Related to the Polyamorphism

In some works, the role of solvent tends to be neglected or simply considered by a mean field-like treatment.

Here we will examine the structural change in the water with salt concentration observed in Chapter 8 more closely with considering the existence of polyamorphism in water.

Since the changes are remarkable for the HW–OW and OW–OW pairs, three systems I-A, II-C, and I-C after gelation were compared in Fig. 9.11. Functions for I-A and II-C are similar. For the system I-C(gel), the sharpening of peaks is observed for both functions for OW-OW and HW-OW pairs. The shoulder at around 3.3 Å becomes clear in this system I-C for the OW-OW pair, which is a typical one for high density water.

During the spontaneous formation of the gel in the I-C system, a gradual shrinkage (approximately 19% in volume) of the system in *NPT* conditions was observed. This volume change is accompanied with the arrest of the dynamics of the system, including the water accompanied with the formation of three-dimensional (infinitive) networks of silica. The pair correlation functions, $g(r)$ of each pair of HW–HW, HW–OW, and OW–OW for the I-C and II-A systems are also shown in Fig. 8.10 in Chapter 8. Although the change observed here is non-equilibrated one, this phenomenon might be related to a transition of the water to the high-density water-like structure, since the shrinkage accompanied with gelation is just expected from the changes of LDL to HDL of water found experimentally in several conditions. Similar structural changes are observed in MD simulations of water–NaCl systems [33–36] related with shoulders at around 3.3 Å in OW–OW structures.

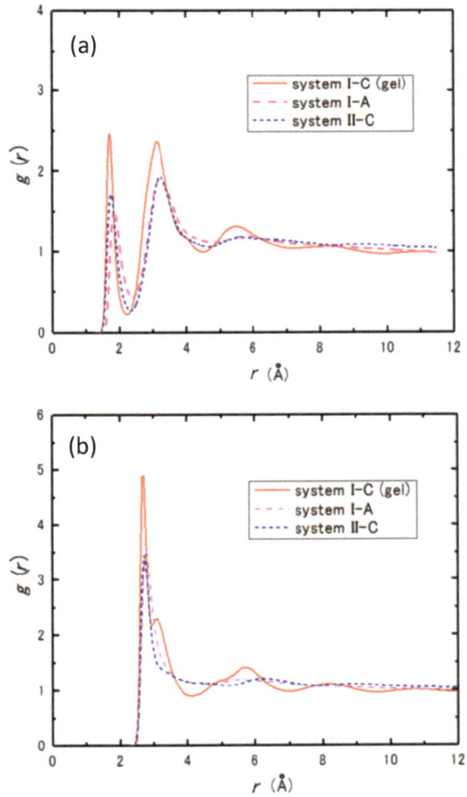


**Figure 9.11** Pair correlation functions of water in three systems are compared for (a) HW–OW pairs and (b) OW–OW pairs. Changes in the peak heights are clear. With increasing concentration of NaCl, the shoulder of the first peak for OW–OW pairs develops at around 3.3 Å. After gelation, peaks become sharper and long-range oscillation appears with trapping of water.

The effect of the addition of the salt has been examined by both experiments and simulations. Mancinelli and coworkers have shown the structure of LDL and HDL affected by monovalent salt by neutron diffraction [33]. In the TIP4P model of water, Corradini and coworkers have argued that LDL is more affected by the presence of the salt than in HDL, where the chloride ion tends to substitute the position of oxygen atoms [34]. The latter trend has been also observed in the SPC/E water–NaCl system [36]. It is interesting to note that quite a similar situation is reproduced well [37] by the monatomic water model "mW" [38, 39] as shown in Fig. 9.12. The model represents each molecule as a single particle that interacts through anisotropic short-ranged potentials and the potential favor "hydrogen-bonded" water structures. The effect of the salt on properties of water was a target of debate for a long time [40, 41]. For example, by Monte Carlo simulations of MB (the model like a logo of Mercedes-Benz), two main ideas are captured in [40]: (1) that charge densities govern the interactions of ions with water and (2) that a balance of forces determines water structure: electrostatics (water's dipole interacting with ions) and hydrogen bonding (water interacting with neighboring waters). Small ions have high charge densities, so they cause strong electrostatic ordering of nearby waters, breaking hydrogen bonds. In contrast, large ions have low charge densities, and surrounding water molecules are largely hydrogen bonded. However, the characteristics of structures found in [33–36] related to the shoulder or sub-peak at ~3.5 Å are not necessarily explained by these concepts. For the argument of polyamorphism of water, the inclusion of changes of sparse regions between the first and second shell is required [14, 40–42]. Furthermore, contributions of further shells are necessary [14], especially for a quantitative discussion of the density of the system.

Some differences exist when we compare the systems with the same NaCl concentration with and without colloidal silica. One is that the first peak of $g(r)$ for OW–OW pair in the colloidal system tends to be sharper with increasing concentration of the salt with the silica part. This is related to the structural arrest during the gelation process. Another is the more development of subpeaks meaning that both charges on silica part and NaCl play important roles in formation of gels. Caution may be required for the discussion of structures of gels due to non-equilibrium characters because the temperature of the system tends to decrease even under the *NPT* condition. Thus, it seems to better to distinguish the terms "*NPT* condition" and "*NPT* ensemble" as in the case of the treatment of the glass transition problem [43].

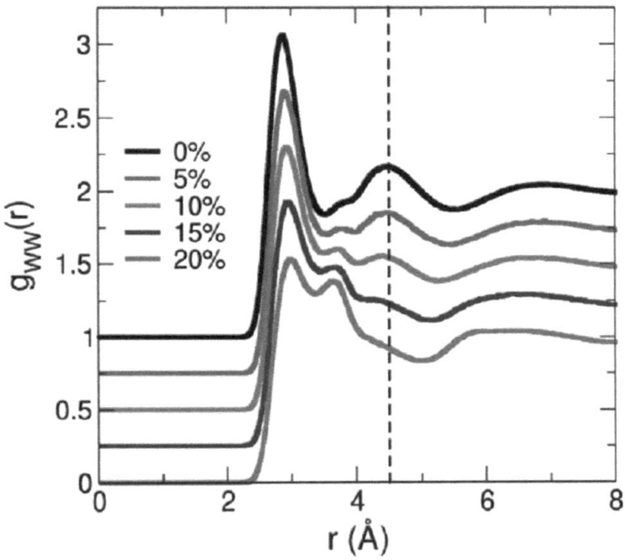


**Figure 9.12** Water–water radial distribution function in the liquid at 300 K in the presence of different molar percentage of solute S (labeled within the figure). The radial distribution functions are vertically displaced to facilitate the visualization. Reprinted with permission from Le L., Molinero, V. (2011), Nanophase segregation in supercooled aqueous solutions and their glasses driven by the polyamorphism of water, *J. Phys. Chem. A*, 115, 5900–5907. Copyright (2011) American Chemical Society.

## 9.5 Structure of the Salt Part in Gels

In some works, the structure of the salt is neglected. Even in MD simulations of aqueous solutions, quite a small amount of the salt tends to be included because many numbers of water molecules are required to treat solutions. As a result, the structures among salt were also neglected. However, when one considered the effects of the salt in coagulations, the development of the NaCl networks with an increasing concentration is a natural expectation. In our approach, the structure of the salt part was also considered. A development of the silica network should be accompanied with this change. In Fig. 9.13 we have visualized the development of the network of the salt part when the gel was formed. If we consider the first coordination shells only, the network formation of salt is not so obvious; however, if we include the second coordination shells, clearer structures of Na–Cl networks are found. Therefore, the examination of length scales of ordering in the NaCl part [36] will be an important step to characterize such structures. The network formation by the addition of the salt is natural since the limiting case with the concentrated NaCl solution can be a molten salt as shown in Chapter 4, where the long-range charge ordering is found [31] and a comparable situation is found in ionic liquids [44].

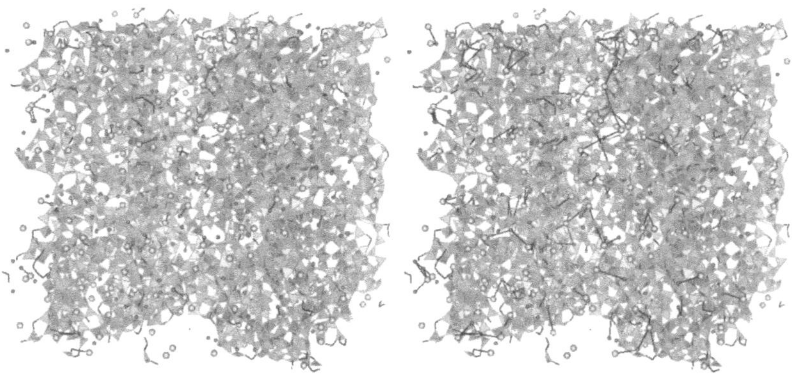

**Figure 9.13** Formation of Na–Cl networks in the system I-C after gelation. (Left) Position of Na and Cl ions in the example shown in Fig. 9.1. The cut-off length $d$ = 3.58 Å was used to show the ionic interaction among Na and Cl. (Right) The same structure, where the cut-off length $d$ = 6.00 Å is used. In the NaCl-rich region, the formation of clusters or chains is clearly found when the second coordination shells were considered. Na is shown in green color and Cl is shown in yellow color. Networks formed by $SiO_4$ polyhedra are shown in gray.

## 9.6 Conclusion

In this chapter, the analyses of the spontaneously formed structures have been done for systems of the colloidal silica–water–NaCl obtained by full atomistic MD simulations. The structures of clusters, aggregates, and gels are compared by several methods. The results clearly show the systematic changes of structures with coagulations and the existence of multiple-networks in the gel is characterized as summarized below.

To learn the structure formation of gels, it is useful to consider individual networks by silica parts, water parts, and salt parts separately and then mixing of them. Several methodologies applied to characterize such networks are shown in this chapter.

From the examination of Si–(O–)Si networks in the gel, the existence of "super-aggregates" (aggregates of aggregates) was confirmed. It is found that the structure is formed by the rearrangement of bonds with keeping the number of bonds and $Q_n$ structures to form the super-aggregates.

In the silica colloids, solvent and salt are incorporated into the fractal structure of the system and hence fractal dimension analyses are useful to characterize these systems. As shown here, the fractal dimension analyses of structures with different magnitude of coagulations are done by both 2-D and 3-D methods and it is found that the 3-D methods are more reliable to distinguish these systems. These values are used to show the existence of super-aggregates in gels.

These structures of clusters or aggregates are separated by solvent-rich regions. The concept of the potential of the mean force is used to understand an effective barrier. That is, the barrier height as a function of the pair correlation function of Si–Si pairs changes with variations in the NaCl concentration and with the silica content. The structure of Si–Si pairs is not determined by other colloids only but by interactions with all other atoms.

With the formation of aggregations and/or gels, the structural change of water is also observed. That is, the formation of smaller domains of water networks is caused by the addition of the salt and it is accompanied with the slowing down of the dynamics. In the water part, structural changes are accompanied with an increase in the density. Such structural changes of water are closely related to its polyamorphism. Thereby, structural changes of solvent need to be treated carefully ions in the coagulation processes.

Na and Cl ions in water are also part of the structure, where they also form networks as naturally expected. Therefore, these systems consist of multiple networks of silica, water, and NaCl. This result suggests the necessity of the inclusion of structures of water and salt to understand the structural formation of aggregates and gels. This picture is not necessarily the same as the conventional explanation by the DLVO model.

As shown in this chapter, atomistic simulations are useful to understand complex systems such as gel, where the multifractal nature of the structure is presented. At the same time, the importance of the inclusion of long-ranged structures is pointed out. Since multifractal structures are widely observed in nature, the argument of the structures of gels in this chapter will be useful to understand many other systems including nanostructured materials. One can imagine how such long-ranged structures are modified by nano-sizing. In nanostructured materials, not only the surface area but also the changes in length scales should play important roles. MD simulations are useful to examine details of such situations.

## References

1. Habasaki, J., Ishikawa, M. (2014), Molecular dynamics study of coagulation in silica-nanocolloid–water–NaCl systems based on the atomistic model, *Phys. Chem. Chem. Phys.,* 16, pp. 24000–24017.
2. Hasmy, A., Anglaret, E., Foret, M., Pelous, J., Jullien, R. (1994), Small-angle neutron-scattering investigation of long-range correlations in silica aerogels: Simulations and experiments, *Phys. Rev. B,* 50, pp. 6006–6016.
3. Kirkwood, J. G. (1935), Statistical mechanics of fluid mixtures, *J. Chem. Phys.,* 3, pp. 300–313.
4. Brandriss, M. E., Stebbins, J. F. (1988), Effects of temperature on the structures of silicate liquids: $^{29}$Si NMR results, *Geochimi. Cosmochim. Acta,* 52, pp. 2659–2669.
5. Henderson, G. S., Calas, G., Stebbins, J. F. (2006), The structure of silicate glasses and melts, *Elements,* 2(5), pp. 269–273.
6. Maekawa, H., Maekawa, T., Kawamura, K., Yokokawa, T. (1991), The structural groups of alkali silicate glasses determined from 29Si MAS-NMR, *J. Non-Cryst. Solids,* 127, pp. 53–64.
7. Greaves, G. N., Sen, S. (2007), Inorganic glasses, glass-forming liquids and amorphizing solids, *Adv. Phys.,* 56, pp. 1–166.

8. Doweidar, H. (1996), The density of alkali silicate glasses in relation to the microstructure, *J. Non-Cryst. Solids,* 194, pp. 155–162. https://doi.org/10.1016/0022-3093(95)00489-0.

9. Habasaki, J., Ngai, K. L. (2013), Molecular dynamics study of network statistics in lithium disilicate: $Q_n$ distribution and the pressure-volume diagram, *J. Chem. Phys.*, 139, 064503 (1–14).

10. Stebbins, J. F., McMillan, P. (1989), Five- and six-coordinated Si in $K_2Si_4O_9$ glass quenched from 1.9 GPa and 1200 °C, *Am. Mineral,* 74(7), pp. 965–968.

11. Mishima, O., Stanley, H. E. (1998), The relationship between liquid, supercooled and glassy water, *Nature,* 396, pp. 329–335.

12. Swamy, V., Saxena, S. K., Sundman, B., Zhang, J. (1994), A thermodynamic assessment of silica phase diagram, *JGR Solid Earth*, pp. 11787–11794.

13. Poole, P. H., Hemmati, M., Angell, C. A. (1997), Comparison of thermodynamic properties of simulated liquid silica and water, *Phys. Rev. Lett.* 79, 2281–2284.

14. Habasaki, J. (2019), Heterogeneous–homogeneous transition and anomaly of density in SPC/E water examined by molecular dynamics simulations, *Physica A*, 527, 121391.

15. Mandelbrot, B. (1967), How long is the coast of Britain? statistical self-similarity and fractional dimension, *Science*, 156, pp. 636–638.

16. Ben-Avraham, D., Havlin, S. (2000), *Diffusion and Reactions in Fractals and Disordered Systems*, Cambridge University Press, Cambridge.

17. Honda, K., Toyoki, H., Matsushita, M. (1986), A theory of fractal dimensionality for generalized diffusion-limited aggregation, *J. Phys. Soc. Jpn.,* 55, p. 707–710.

18. Meakin, P. (1987), Fractal aggregates, *Adv. Colloid Interface Sci.,* 28, pp. 249–331.

19. Fry, D., Chakrabatti, A., Kim, W., Sorensen, C. M. (2004), Structural crossover in dense irreversibly aggregating particulate systems, *Phys. Rev. E*, 69, 061401(1–10).

20. Allen, M. P., Tildesley, D. J. (1989), *Computer Simulation of Liquids*, Oxford University Press, New York.

21. Ueda, A. (2003), *Molecular Simulations: From Classical to Quantum Method*, SHOKABO, Tokyo, in Japanese.

22. Stauffer, D. (1985), *Introduction to Percolation Theory*, Taylor & Francis, London and Philadelphia, and references therein.

23. Weitz, D. A., Huang, J. S., Lin, M. Y., Sung, J. (1985), Limits of the fractal dimension for irreversible kinetic aggregation of gold colloids, *Phys. Rev. Lett.*, 54, 1416(1–4).

24. Lin, M. Y., Lindsay, H. M., Weitz, D. A., Ball, R. C., Klein, R., Meakin, P. (1989), Universality in colloid aggregation, *Nature*, 339, pp. 360–362.

25. Schaefer, D. W., Martin, J. E., Wiltzius, P., Cannell, D. S. (1984), Fractal geometry of colloidal aggregates, *Phys. Rev. Lett.*, 52, 2371(1–5).

26. Mandelbrot, B. B. (1999), *Multifractals and 1/f Noise: Wild Self-Affinity in Physics (1963–1976)*, Selecta Volume N. Springer.

27. Chhabra, A., Jensen, R. V. (1989), Direct determination of the f(alpha) singularity spectrum, *Phys. Rev. Lett.*, 62, pp. 1327–1330.

28. Halsey, T. C., Jensen, M. H., Kadanoff, L. P., Procaccia, I., Shraiman, B. (1986), Fractal measure and their singularities: The characterization of strange sets, *Phys. Rev. A*, 33, pp. 1141–1151. doi:10.1103/PhysRevA.33.1141.

29. Sorensen, C. M., Hageman, W. B., Rush, T. J., Huang, H., Oh, C. (1998), Aerogelation in a flame soot aerosol, *Phys. Rev. Lett.*, 80, 1782(1–4).

30. Mishima, O., Calvert, L. D., Whalley, E. (1984), 'Melting ice' I at 77 K and 10 kbar: A new method of making amorphous solids, *Nature*, 310, pp. 393–395.

31. Poole, P. H., Sciortino, F., Essmann, U., Stanley, H. E. (1992), Phase behavior of metastable water, *Nature*, 360, pp. 324–328. doi:10.1038/360324a0.

32. Soper, A. K., Ricci, M. A. (2000), Structures of high-density and low-density water, *Phys. Rev. Lett.*, 84, pp. 2881–2884.

33. Mancinelli, R., Botti, A., Bruni, F., Ricci, M. A., Soper, A. K. (2007), Perturbation of water structure due to monovalent ions in solution, *Phys. Chem. Chem. Phys.*, 9, pp. 2959–2967.

34. Corradini, D., Rovere, M., Gallo, P. (2011), Structural properties of high and low density water in a supercooled aqueous solution of salt, *J. Phys. Chem. B*, 115, pp. 1461–1468.

35. Le, L., Molinero, V. (2011), Nanophase Segregation in supercooled aqueous solutions and their glasses driven by the polyamorphism of water, *J. Phys. Chem. A*, 115, pp. 5900–5907.

36. Gubarevich, A., Habasaki, J. (2017), Long-range charge order and electrostatic screening in ionic liquids and concentrated electrolyte solutions, *8th International Discussion Meeting on Relaxations in Complex Systems (IDMRCS8)*, Wisla, Poland.

37. Molinero, V., Moore, E. B. (2009), Water modeled as an intermediate element between carbon and silicon, *J. Phys. Chem. B,* 113, pp. 4008–4016.

38. Ben-Naim, A. (1971), Statistical mechanics of "waterlike" particles in two dimensions. I. Physical model and application of the percus–yevick equation, *J. Chem. Phys.*, 54, pp. 3682–3695.

39. Urbic, T., Ken Dill, A. (2010), A statistical mechanical theory for a two-dimensional model of water, *J. Chem. Phys.,* 132, 224507(1–9).

40. Cox, W. M., Wolfenden, J. H., (1934), The viscosity of strong electrolytes measured by a differential method, *R. Soc. London Ser. A,* 145, pp. 475–488.

39. Hribar, B., Noel Southall, T., Vlachy, V., Dill, K. A. (2002), How ions affect the structure of water, *J. Am. Chem. Soc.*, 124(41), pp. 12302–12311.

40. Bandyopadhyay, D., Mohan, S., Ghosh, S. K., Choudhury, N. (2013), Correlation of structural order, anomalous density, and hydrogen bonding network of liquid water, *J. Phys. Chem. B,* 117, pp. 8831–8843.

41. Matsumoto, M. (2009), Why does water expand when it cools?, *Phys. Rev. Lett.,* 103, 017801 (1–4).

42. Székely, E., Varga, I. K., Baranyai, A. (2016), Tetrahedrality and hydrogen bonds in water, *J. Chem. Phys.,* 144, 224502 (1–10).

43. Habasaki, J., Ueda, A. (2016), Molecular dynamics study of one component soft-core system—Analytical expression of non-equilibrium relaxation in constant pressure conditions, *J. Non-Cryst. Solids*, 447, pp. 212–222.

44. Habasaki, J., Ngai, K. L. (2011), Multifractal nature of heterogeneous dynamics and structures in glass forming ionic liquids, *J. Non-Cryst. Solids,* 357, pp. 446–453.

# Afterword

This book discusses historical view and the current status of the field of nanoionics, nanostructured materials, and nanotechnologies from the nanoscopic point of view based on molecular dynamics simulations. After general discussion for classifications by dimensionality of systems and effects of nanosizing, the systems treated are expanded to disordered systems such as nanoporous glasses and gels.

Dynamics and structures are characterized by several methods, such as length scales in charge density wave, fractal dimensions including the case multifractals, and $Q_n$ distributions in networks.

One of the key problems treated in this book is to examine the mechanism of enhanced dynamics of ions found in porous lithium silicates. Similar problems are found in porous composites and/or films, for which many theories and models have been proposed so far. How can we distinguish possible origins of the enhancement in porous lithium silicate systems? In our results for porous lithium disilicate systems examined by molecular dynamics, the importance of caged ion dynamics and the percolative aspect at longer time and length scales are clearly shown with distinguishing different length scales and time scales. Essential parts of the method used will be applicable for other systems showing enhancement. Porous lithium silicate systems also show self-organization of larger voids by the expansion of the system under the constant volume conditions and also show self-healing processes in the following runs under constant pressure conditions. Changes in the structures of coordination polyhedra and pressures gave insights for the mechanism of structural formation, mechanical properties, and dynamics in both processes.

Another problem treated in this book is concerned with the coagulation process of the colloidal solutions with salt examined by full atomistic model, although such treatment is challenging due to large size difference in colloids and solvents. The existence of "superaggregtion," which means complex structures having more than one fractal dimension, is clearly found in the structure of gel, spontaneously formed during fully atomistic MD simulations. The result of MD simulations revealed that the water structure changed with the concentration of salt, and this suggests the necessity of the inclusion of structures of water for the treatment of coagulation processes, especially in nanocolloidal systems. Interestingly, the structural change of water is closely related to the problem of the polyamorphism.

Both enhancement of ionic motions in porous systems and gelation in colloidal systems mentioned above are concerned with remarkable changes in dynamics, where dynamic heterogeneity is commonly observed. The existence of multifractality (coexistence of different length scales) is a key feature to characterize them and several tools related to them are introduced to treat them.

I would like to call readers' attention for common behaviors of systems, which will result in the further development of related fields. These concepts, such as dynamic heterogeneity and related multifractality, are rather general [1, 2] and suggest the existence of the common underlying physics. Understanding it is useful for both fundamental physics and applications. For example, modification of dynamics in porous materials is related to the stabilization of drugs in the drug delivery system. Thus both porous materials and colloidal systems grow in importance in many fields, including biotechnology [3, 4], environmental chemistry [5–7], pharmaceuticals [8], and industry. Of course, such developments including other nanostructured materials cannot be separated from the development of experimental techniques for characterization and synthesis methods [9] as well as theoretical methods and simulations.

As shown in this book, MD simulations can play crucial roles in understanding fundamental mechanisms and have potentials to develop new functional nanostructured materials. I hope readers will have fabulous ideas of how to use MD simulations in these fruitful fields. These systems are useful not only for applications but also as platforms [10] to study common physics such as the problem of the glass transition.

## References


1. Habasaki, J., León, C, Ngai, K. L. (2017), *Dynamics of Glassy, Crystalline and Liquid Ionic Conductors: Experiments, Theories, Simulations*, Springer International, Switzerland.
2. Ngai, K. L. (2011), *Relaxation and Diffusion in Complex Systems*, Springer, New York.
3. Tadros, T. (2014), Colloid and interface aspects of pharmaceutical science, in Ohshima, H. Makino, K. (eds.), *Colloid and Interface Science in Pharmaceutical Research and Development*, Elsevier, chapter 2, pp. 29–54.
4. Milcovich, G., Lettieri, S., Antunes, F. E., Medronho, B., Fonseca, A. C., Coelho, J. F. J., Marizza, P., Perrone, F., Farra, R., Dapas, B., Grassi, G., Grassi, M., Giordani, S., (2017), Recent advances in smart biotechnology: Hydrogels and nanocarriers for tailored bioactive molecules depot, *Adv. Colloid Interface Sci.*, 249, pp. 163–180.
5. Mahmood, Q., Khan, A. F., Khan, A. (2014), Colloids in the environmental protection—current and future trends, in Fanun, M., ed., *The Role of Colloidal Systems in Environmental Protection*, Elsevier, chapter 25, pp. 635–677.
6. Ma, J., Qin, G., Zhang, Y., Sun, J., Wang, S., Jiang, L., Heavy metal removal from aqueous solutions by calcium silicate powder from waste coal fly-ash, *J. Clean. Prod.*, 182, 2018, pp. 776–782.
7. Alothman, Z. A. (2012), A review: Fundamental aspects of silicate mesoporous materials, *Materials*, **5**(12), pp. 2874–2902, doi:10.3390/ma5122874.
8. Kamaly, N., Yameen, B., Wu, J., and Farokhzad, O. C. (2016), Degradable controlled-release polymers and polymeric nanoparticles: Mechanisms of controlling drug release, *Chem. Rev.*, 116, 4, pp. 2602–2663.
9. D. Brabazon, D. (2016), *Nanostructured Materials, Reference Module in Materials Science and Materials Engineering*, Elsevier, ISBN 9780128035818, https://doi.org/10.1016/B978-0-12-803581-8.04103-5.
10. Habasaki, J., Ngai, K. L. (2018), Heterogeneous dynamics in nano porous materials examined by molecular dynamics simulations: Effects of modification of caged ion dynamics, *J. Non-Cryst. Solid.*, 498, pp. 364–371.

# Index